A Simple Guide to Telescopes, Spotting Scopes & Binoculars

Bill Corbett

Contents

First published in the United States in 2003 by
Watson-Guptill Publications,
a division of VNU Business Media, Inc.
770 Broadway, New York, NY 10003
www.watsonguptill.com

Author: Bill Corbett
Editor: Jenny Corbett
Design: Bill Corbett
Pre-press Production: Everbest Printing Co., Ltd.
Printed & Bound: Everbest Printing Co., Ltd.

Library of Congress Cataloging-in-Publication Data:
Corbett, Bill, 1948-
  A simple guide to telescopes, spotting scopes & binoculars / Bill Corbett.
    p. cm.
    Includes index.
  ISBN 0-8174-5888-3
  1. Telescopes. 2. Binoculars. I. Title: Simple guide to telescopes, spotting scopes, and binoculars. II. Title.
  QB88.C775 2003
  522'.2–dc21
                                        2003008394

Printed in China
1 2 3 4 5 6 7 8/09 08 07 06 05 04 03

Introduction

Thank you for purchasing this copy of "A Simple Guide to Telescopes, Spotting Scopes and Binoculars."

You will find that this book is very reader-friendly and easy-to-understand.

A word of warning regarding the three optical products this book discusses.

When using them **never look directly at the Sun without appropriate solar filters attached to the lenses.** Immediate <u>damage to your eyes</u> and equipment can result if this advice is ignored.

A bit of helpful advice right at the start. If you are going to purchase a telescope, spotting scope or binoculars, buy from someone who understands the product, and preferably uses it.

This especially applies to telescopes.

If you are going to get into telescopes get in contact with an astronomy club in your area before you rush out and make a purchase.

They quite often have meetings and open nights including "star parties" where newcomers are welcome and their equipment is on display to see and use.

Talk to them and get their advice, read this book and then you should be in a much better position to make an informed purchase.

Also please be advised that this book does not go into the intricate details of, say, how to find M3 globular cluster or other celestial bodies, using a telescope.

If you need that sort of information you should look for a specialized book on the subject.

What this book can do is to get you started with as few problems and hassles as possible in regard to the purchase, setting up, use and care of your new equipment.

It also gives basic details of how to get started and find objects in the sky, if you are using a telescope.

The book is not comprehensive in this area but will certainly get you going and pointed in the right direction.

Once that is done there are numerous books and computer programs that can take you further in your exploration of space.

In the past I have concentrated my publishing endeavors on photography which, of course, put me in contact with many retail camera and telescope outlets.

Many camera stores that I deal with informed me that there was a lack of easy-to-understand reading material on the telescopes and binoculars available on the market.

Subsequent investigation confirmed this to be true.

Another astounding fact came to my attention. Even some of the manufactures of binoculars, spotting scopes and telescopes seem to have an aversion to printing information about their products that would be of assistance to their customers.

The number of products that I saw that came with no instructions or, at best, minimal or confusing information was truly amazing.

You would think that if a customer was prepared to part with their hard-earned money to purchase a product then the manufacturer, at the very least, had the responsibility to explain how to properly assemble (if needed), care and use the product in a manner that was simple and easy to understand.

Not so in many cases I'm afraid!

Even binoculars, which are pretty simple products to use, can benefit from some explanation on their use and care, as well as on what to look out for when purchasing them.

Having got all the above off my chest what can you expect from this book?

For a start, it's a good idea to understand how a particular piece of equipment works.

Next we will look at the three classes of products that this book covers and go into some detail on the following points:

Binoculars

- Binocular types and applications
- Correct use of binoculars
- Binocular care and maintenance
- What to look for when purchasing binoculars

Canon 15 x 50 IS All Weather binoculars

Spotting Scopes

- Spotting scopes and their applications
- Setting up and correct use of spotting scopes
- Care and maintenance of spotting scopes
- What to look for when purchasing a spotting scope

Fujinon spotting scope on tripod

Telescopes

- Telescope types (refractive, reflective and catadioptrics) and their applications
- Assembling your telescope
- How to set up and adjust your telescope correctly
- Using a telescope
- Telescope care and maintenance
- What to look for when purchasing a telescope.

A Celestron Schmidt-Cassegrain telescope with computer control facilities

So for a relatively small volume we are going to cover a fair bit of territory. But we will do it simply and in as interesting a manner as possible.

Care and maintenance of binoculars, spotting scopes and telescopes is covered individually later in this book, as mentioned above, but there is one point I would like to stress right at the start.

This covers not only binoculars, spotting scopes and telescopes but also camera lenses, microscopes and so on.

I asked the manager of one of Australia's largest specialty optical retail outlets how much shock telescopes, binoculars and so on could absorb, without being damaged.

His answer was as follows: "The biggest shock an optical product should endure is for you to shout 'boo' when you open the box."

Treat your equipment carefully, even if it is labelled "Shock Proof"!

One final thing. Telescopes often give vital measurements in either inches or millimeters, and sometimes both.

It can be helpful to remember that one inch equals 25.4 mm. Therefore, if you have a telescope that specifies it has a 2-inch objective lens simply multiply the number of inches by 25.4 to convert it to millimeters.

Obviously, in the example given, the diameter of this objective lens, specified in inches, would equal 50.8 mm (2 in. x 25.4 = 50.8 mm.)

To convert millimeters to inches just do the reverse and divide the number of millimeters by 25.4 to arrive at the number of inches.

I hope you enjoy this book and that it assists you in enjoying your equipment and the wonders that it can reveal to the fullest.

Bill Corbett
Sydney, Australia

Star Parties

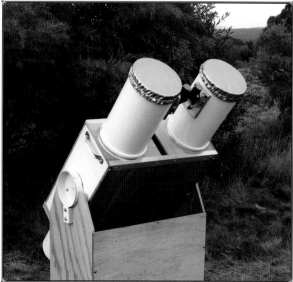
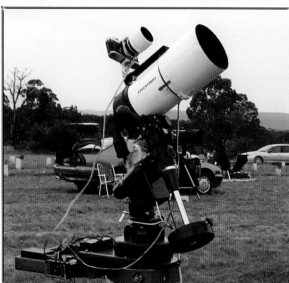

Many astronomy clubs conduct star parties at least once a year. The public is usually welcome at these events, and for those interested in telescopes and binoculars they are most interesting. You will see a great variety of equipment, some commercially made, others homemade. The biggest advantage of star parties is that you can see and try out a great variety of equipment and talk to experienced people about choices that fit your needs, budget and experience.

In addition, many star parties feature guest speakers who shed light on the fascinating subject of astronomy. Last, but not least, they can be a great deal of fun. Just remember to take a red flashlight with you (see Telescope Accessories).

Telescopes: How They Work and What to Look For

Since binoculars and spotting scopes both evolved from the telescope, a short review of its development is called for.

A Short History

A Dutch spectacle maker, Hans Lippershey (1570–1619) is credited with actually inventing the telescope, but there is some controversy regarding this.

What we do know is that in August of 1609 Galileo (1564–1642) exhibited the world's first telescope. It was a basic refractor scope (more on refractors a little later) with around 20x magnification. He improved the design later in the same year.

Galileo's subsequent observations led him to state that the earth moved around the sun, a theory that did not please the Catholic Authorities at the time.

He was subsequently tried and found guilty of "grave suspicion of heresy" and sentenced to life imprisonment, which was later commuted to permanent house arrest.

It was not until 1992 that the Catholic Church reversed its condemnation of Galileo. Obviously they hate to move quickly!

The development of the telescope then continued with a host of innovations that improved performance.

Notable among them was Sir Isaac Newton's invention of the reflector telescope in 1668 (more on reflectors later), the invention of the achromatic object glass by John Dolland in 1757 and the development of the equatorial mount by William Lassell in Britain in the 18th century.

As with all things, the development of telescopes continues today with new and improved models constantly appearing.

A relatively recent development, that certainly can make life easier for the budding astronomer, as far as finding your way around the heavens, is the incorporation of both computer chips and Global Positioning System (GPS) technology with telescopes.

These new models can make the location of distant objects much easier, although at a price. More on this later.

Please note that we will be looking only at the models that are available to the consumer, so don't expect a dissertation on the Hubble or the Very Large Array radio telescope.

Telescopes

There are three types of telescope that are generally available to the public. These are the Refractor, Newtonian Reflector (catoptric) and the Catadioptric telescopes.

The catadioptrics then divide into two types known as the Schmidt-Cassegrain and the Maksutov-Cassegrain.

We will look at the differences between the two kinds.

All these different scopes have the same purpose in life, which is to capture light being reflected from an object and bring that light to the focal point where it can be viewed, with the aid of an eyepiece.

Now, while we are speaking about light, let's digress for a moment and discuss a very important feature of all types of telescopes. That feature is the aperture.

The aperture controls how much light the telescope is capable of picking up and <u>this is the single most important feature of a telescope</u>.

Obviously, the more light it can collect, the brighter and clearer the image. As with camera lenses, the bigger the aperture of a telescope the more light it is capable of collecting.

So in the case of telescopes, *bigger is better,* as far as the aperture is concerned!

When it comes to telescopes, the aperture size is normally expressed in inches (sometimes in millimeters) and refers to the diameter of the <u>objective lens</u> with refractors and the <u>primary mirror</u> with other designs of telescopes, such as catoptrics and catadioptrics.

An important tip when making a purchase: Beware of telescopes that are advertised by their magnification factor.

The eyepiece views the image being projected by either the objective lens or primary mirror and magnifies the projection so the observer can see.

If the aperture is small, then the light collected by the objective lens or primary mirror is less than what it could be. No amount of magnification will make up for this inability to gather light caused by the small size of the aperture, or inferior optics.

Okay, the aperture is important. But apart from this, how do telescopes work? What are the differences, and what are the pros and cons for each type?

Well, let's have a look and see if we can sort out all this in detail.

The Refractor Telescope (Sometimes Called a Dioptric)

Since it was the first to be invented, let's start with the refractor telescope. It is also the design that most people associate with the word *telescope.*

A refracting telescope is one that comprises a long tube. At the front of the tube is a lens called the objective lens, whose role is to gather light being reflected from the object being observed and focus it to the rear of the tube, at the focal point.

At the rear of the tube is the eyepiece, which is another type of lens used to both magnify and view the image of the distant object, created at the focal point by the objective lens.

The eyepiece, which is interchangeable with other eyepieces providing different levels of magnification, sits in a collar. By turning a knob that moves the collar, you can focus on the image at the focal point.

Since many telescopes are pointed at the night sky, this can mean that the eyepiece is pointing down low to the ground.

To correct both problems (of the images being reversed and upside-down) in a refractor telescope you need to use a prism with your eyepiece or, alternatively, a star diagonal with an erecting lens.

The diagram below shows how refractors work.

Objective Lens

Eyepiece

The light path of a refractor telescope.

Using a telescope this way can be pretty uncomfortable and produce neck strain.

To counter this, refractor telescopes can generally be fitted with what's known as a star diagonal. This allows the eyepiece to point upwards, a far more comfortable viewing position.

In addition, if the eyepiece is directly at the back of the tube, then the image you see will be upside-down and mirror-reversed.

Now, if you are observing some distant star or planet, then the fact that your image is upside-down and mirror-reversed is not important.

But if you are viewing some distant object on earth, then alignment does become important and the use of the star diagonal will correct the image so that it is no longer upside-down.

However, it will still be mirror-reversed.

Refractor telescopes can use two types of objective lens.

The first is known as an achromatic refractor design. This type uses two lens elements to help reduce what's known as *chromatic aberration,* which is an optical effect that causes the different wavelengths of

A refractor telescope fitted with a star diagonal and mounted on a tripod with an altazimuth mount.

light, passing through the lens, to focus at slightly different points.

The second design is called apochromatic refractors. This is where the objective lens is comprised of three or more lens elements to completely eliminate chromatic aberration.

Advantages of the Refractor Telescope

1   Because it has a simple design, it is very reliable and easy to set up and use.

2   The fact that the objective lens is permanently mounted and aligned is a very big plus.

3   Very little needs to be done to maintain the refractor telescope, other than to clean the body, objective lens and eyepiece.

4   It produces images of high contrast, as there is no secondary mirror or other obstruction in the light path.

5   They are ideal for long-distance terrestrial viewing.

6   Because it is, in effect, a sealed tube, internal air currents, which can affect you image quality, are greatly reduced.

7   The use of apochromatic, fluorite and ED design with the lens ensures excellent color correction. Even where achromatic lens designs are used, color correction is still good.

8   The refractor telescope is excellent for observing such objects as the moon, planets and binary stars.

Disadvantages of the Refractor Telescope

1   The cost of the refractor telescope is greater than the Newtonian or catadioptric telescope when you compare the equivalent aperture size.

2   Design limitations and costs constrain the size of the aperture of the telescope. This limits the suitability of using the refractor telescope to both view and photograph faint astronomical bodies.

3   Again the design makes refractor telescopes physically bigger than equivalent sized (in relation to aperture) Newtonian or catadioptrics with the attendant problem that they will weigh more.

4   Refractor telescopes sometimes have a poor reputation. This, in many cases, is caused by the poor quality of cheap toy refractors that abound in the marketplace. A high quality refractor telescope is a fine instrument and should not be compared to the cheaper models.

The next telescope type we will look at is the reflector telescope which was invented by Sir Isaac Newton and sometimes called the Newtonian Reflector.

Reflector Telescope (Sometimes Called the Newtonian Reflector or Catoptric)

Unlike the refractor model, the reflector telescope has no lens at the front of the tube. It is also greater in diameter than a refractor model.

Instead of the lens, the Newtonian reflector employs a concave objective mirror at the back of its tube.

Towards the front of the tube is a much smaller flat secondary mirror, positioned in the center of the tube, whose job it is to divert the reflected light rays to an eyepiece for viewing.

The eyepiece is positioned at the front of the tube at a right angle to the secondary flat mirror. The image will be upside-down but not mirror-reversed.

The light travels down the end of the tube and then is bounced back to the secondary

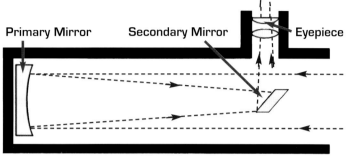

mirror, which in turn diverts the light to the eyepiece.

The light path of a reflector telescope.

Because the reflector telescope employs a mirror for gathering light, it can present a very bright image to the observer.

Looking down the tube of a reflector telescope. The primary mirror can be seen at the bottom of the tube and this also shows the reflection of the secondary mirror and the spider that secures it .

A variation of the reflector telescope is the Dobsonian telescope, which is a Newtonian reflector on a special mount that rests on the ground.

The Dobsonian mount has an altazimuth design, whose movements are constricted

A reflector telescope on an equatorial mount with tripod

to lateral and vertical (see Telescope Accessories). This design is well-suited to large Newtonian telescopes and makes them much easier to use.

Dobsonian telescopes are sometimes called "light buckets" and are many people's choice for their first telescope.

A Dobsonian telescope is considered by many as an ideal first choice when starting up in astronomy.

Advantages of the Reflector Telescope

1   These types of telescopes are good for looking at the planets and moon.

2   Because they have a fast focal ratio they are excellent for looking at faint objects in deep space. Focal ratio refers to the telescope's effective f-number (see Glossary).

3   The cost of producing primary mirrors is less than objective lenses, especially with medium and large apertures, therefore the

consumer can get more light-gathering power for his/her dollar.

4   They are fairly compact and can be transported reasonably easily, albeit with a great deal of care.

5   Because they have fairly fast focal ratios (f-stops) they are capable of delivering bright images that have low optical aberrations and shorter exposure times for astrophotography.

Disadvantages of the Reflector Telescope

1   This type of telescope is not very suitable for terrestrial viewing as the image is inverted. The use of a star diagonal to correct this would be counterproductive.

2   In comparison with a refractor telescope of equivalent aperture, it naturally exhibits a loss of light due to the secondary mirror. However, the loss of light is minimal and the larger and faster apertures more than compensate for this.

3   They are less convenient and more difficult to use for deep space observations compared to cata-dioptrics, which are discussed next.

4   Reflectors with aperture diameters over 8 inches are expensive and tend to be weighty and bulky.

5   Due to their design, they tend to suffer from what's known as off-axis

coma. This can produce distortion of the image at the edge of the field of view. That said, with a properly designed and manufactured telescope, the image in the center of the field of view will be distortion-free.

6   Reflector telescopes tend to be more prone to require maintenance such as collimation, a task that many people find difficult. Collimation is covered later in this book.

7   Because reflector telescopes have an open-tube design, contaminants in the air get in and, over a period of time, degrade the primary and secondary mirror coatings, which affects the telescope's performance.

In addition, the same open-ended design allows air currents to circulate in the tube, which can also affect image quality. For optimum performance, the internal temperature of the air in the tube should be the same as the external air temperature.

Catadioptric Telescopes (Schmidt-Cassegrain and Maksutov-Cassegrain)

Catadioptric telescopes are an evolution of the Newtonian reflector design. Where the Newtonian uses two mirrors, these telescopes use two mirrors and a special lens, sometimes called a corrector lens, to bend the light and deliver it to the focal point at the eyepiece.

The difference between the Schmidt-Cassegrain and the Maksutov-Cassegrain is in the design of this corrector lens.

The Schmidt-Cassegrain Telescope

With this type of telescope, the light enters the telescope through the corrector lens, which is at the front of the tube.

It travels to the primary mirror positioned at the bottom of the tube and is reflected back to the front where the secondary mirror is positioned on the inside face of the corrector lens.

The light is again reflected to the back of the tube to a hole in the center of the primary mirror where the focal point and eyepiece are positioned.

A Schmidt-Cassegrain telescope with equatorial mount

13

Because the light travels the length of the tube three times before reaching the focal point and the eyepiece, the focal length (see Glossary) of the telescope is quite long, although the physical telescope is compact.

Schmidt-Cassegrain telescopes normally have an aperture of around f10, although some models claim to have faster specifications.

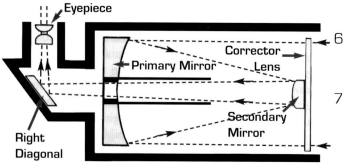

The light path of a Schmidt-Cassegrain telescope.

These models are normally fitted with a focal reducer or corrector which reduces their effective f-stop number.

Advantages of the Schmidt-Cassegrain Telescope

1 This design combines all the advantages of both the lens and mirror models and cancels the disadvantages of each. Many consider them to be the best consumer telescope available.

2 The design produces the best near-focus capability of any telescope type.

3 Because of the closed-tube design, they are durable and maintenance is reduced, but still necessary. Also degradation of mirror surfaces due to air contamination is eliminated.

4 Schmidt-Cassegrain telescopes have very good optical characteristics and produce images that are very sharp, over a wide field of view.

5 Their design makes them very compact and easy to move around, with the appropriate amount of care!

6 They are very versatile, durable and easy-to-use.

7 They are adaptable and as such excellent for terrestrial, lunar, planetary and deep space observations and photography.

8 Compared to refractor designs at equivalent apertures, they are reasonably priced, which makes large aperture models more affordable.

Disadvantages of the Schmidt-Cassegrain Telescope

1 Schmidt-Cassegrain telescopes are more expensive than Newtonian reflectors of comparable aperture.

2 As with Newtonian reflectors there is a slight light loss that is attributable to the secondary mirror.

The Maksutov-Cassegrain Telescope

The Maksutov-Cassegrain telescope design is virtually identical to the Schmidt-Cassegrain design discussed previously.

The difference lies in the design of the correction lens at the front of the tube, and the secondary mirror, which is slightly smaller than that of the Schmidt-Cassegrain model.

The correction lens on the Maksutov-Cassegrain is of a different design. It is thicker, with a strong curvature, compared to the thin, aspheric correction lens of the Schmidt-Cassegrain telescope.

The secondary mirror, instead of being a true mirror as is the case with the Schmidt-Cassegrain, is smaller and normally is made by having an aluminized spot on the inside face of the corrector lens.

While the Schmidt-Cassegrain telescope typically has an aperture of f10, the Maksutov-Cassegrain type most likely has an aperture of between f13 to f16.

This is due to the differences in the design of the corrector lenses of the two models.

Advantages of the Maksutov-Cassegrain Telescope

It is virtually the same as the Schmidt-Cassegrain model, except that the secondary mirror is smaller. Because of this, it gives the viewing of planetary objects slightly better definition.

Disadvantages of the Maksutov-Cassegrain Telescope

Again, the same as with the Schmidt-Cassegrain telescope, except for three points:

1   Because the corrector lens is thicker on the Maksutov-Cassegrain models, they are heavier.

2   The thickness of the corrector lens also causes the air in the tube to take longer to match the external temperature, a condition required for optimal performance.

This can be especially evident in models with large apertures.

3   Because of their focal length, astrophotography exposure times are longer.

Spotting Scopes

Basically, a spotting scope is a specialized refractor telescope. It too has an objective lens at the front and an eyepiece at the rear of the tube.

In addition, it has a focusing knob somewhere near the eyepiece, and the eyepiece itself may be adjustable to allow a change of magnification power, similar to the zoom lens on a camera.

The eyepiece may be straight or angled and both types normally will have a prism included to insure the image is not mirror-reversed or upside-down.

Some have eyepieces that are on an adjustable ring, allowing them to be rotated to either side for certain applications.

Spotting scopes have a wider field of view than telescopes.

A spotting scope mounted on a tripod.

Normally, a spotting scope is mounted on a tripod, but in fact, it can also be held by hand. However, mounting it on a tripod reduces the tremor that invariably accompanies the use of any optical instrument by hand, thereby producing better results.

The better spotting scopes available on the market are waterproof, dustproof and shock resistant.

Attaching the eyepiece to the spotting scope.

Spotting Scope Applications

The uses of spotting scopes are many and include:

> Military applications
> Hunting
> Bird watching
> Terrestrial observations
> Astronomical observations
> Target spotting for sports, such as archery and shooting
> Marine applications
> Terrestrial and astronomical photography
> Surveying
> Sports, such as hiking and trekking

Spotting Scopes: What to Look for When Purchasing

When purchasing a spotting scope, look through a variety of models and compare them.

Always look at *exactly* the same scene, so you have a valid impression of how each model performs.

Check that the lenses are coated and the potential light loss problem is corrected.

Focus on a scene that includes lettering, perhaps a street sign. Is the lettering crystal clear? It should be. Now move the spotting scope slightly so that the lettering is at the edge of the image circle. Is it still crystal clear?

If not, then the spotting scope could suffer from what's know as chromatic aberration, which is a defect in the lens design affecting the resolution of the image at the edge of the image circle.

Now, a lot of people think this abberation is not important, since it only affects the edges of a scene, and they tend to concentrate on the center of the image circle.

Not so, I'm afraid. A spotting scope with this defect can be tiring to use and even induce headaches. Avoid this defect!

Don't purchase more power than you will need or use. It is tempting to do this, but it is just a waste of money.

It is very important that you think about what you are going to use a spotting scope for before, not after, purchase.

Do you need a spotting scope with a straight eyepiece or one that is angled at 45 degrees?

Will you use the spotting scope in snow, rain or dusty conditions?

If so, you need a good reliable "All Weather" design, especially for marine use.

Compare prices and ensure you are getting value of money.

Spotting Scope (Care and Maintenance)

The care and maintenance of all types of spotting scopes is fairly minimal, but a few points should be remembered.

Avoid rough treatment of the spotting scope. Bumps are not good for them!

Don't store a spotting scope in a car or other place where excessive heat or humidity can build up.

In fact, it is a good idea when storing any optical instrument to include a sachet of silica gel to help soak up any moisture in the air.

When not being used, the lens covers should be fitted to protect the objective lens and eyepieces.

However, before fitting the lens cover, make sure that there is no moisture on the objective lens or eyepiece.

If you fail to check this, there is the risk that fungus will form on the optics, a condition that is definitely not recommended.

When the body of the spotting scope gets wet or dusty, clean it with a soft cloth.

Avoid harsh solvents at all times and only use a mild, diluted detergent to clean it, if necessary, such as when stubborn marks remain.

Keep the lenses clean using only lens tissues and lens-cleaning solution, which are readily available from camera stores. You can also use a lens blower to remove dust.

Do not pour lens-cleaning solution directly onto the surface of the lens. It could penetrate the interior of the scope and give you major problems.

Instead, apply some solution to a lens tissue and wipe the lens surface gently in straight lines, starting at the center and finishing at the edge.

Once this is done, take a clean, dry lens tissue and repeat the wiping procedure to clean up any moisture and dirt until the lens surface is perfectly dry.

If you need to do it a second time use a fresh tissue.

Do not use handkerchiefs and normal tissues to clean lenses, unless the idea of scratches across your lens appeals.

Under no circumstances should you disassemble your spotting scope, unless you have a Ph.D. in optical engineering.

If there is a problem with your spotting scope, take it to a qualified repair center for service.

Telescope care and maintenance is covered in a later chapter.

This archer uses a spotting scope to check the target after each shot. This helps him adjust the accuracy of his next shot. A spotting scope with an eyepiece that adjusts from side to side would keep him from having to move his feet while checking his shot. However, the model he is using does not have this feature. He has compensated for this by using the adjustable tripod head to achieve the same effect.

Telescope Accessories

When it comes to accessories, telescopes are a bit like cameras. They are many and varied.

Some are not necessary, while others are either mandatory or will help you greatly in your observation of either astronomical or terrestrial objects.

For instance, it's fine to have a telescope, but unless you have it mounted correctly on a firm, level foundation, such as a tripod, you are going to find it a bit heavy and cumbersome to use, to say the least.

To attach it to a tripod, you also have to have a mount. What kind of mount you choose is critical when you wish to pursue astrophotography.

The accessories discussed below are either mandatory or helpful in getting better images to either view or photograph, and sometimes both.

Some telescopes will come with some of the accessories mentioned. With others, you will have to purchase them separately.

Tripods

Telescopes are normally large pieces of equipment of considerable weight. Therefore, it is very important that they are mounted in a secure fashion and can be easily adjusted.

You do not want your telescope falling over, do you?

It is also important that telescopes are balanced correctly, and are perfectly level.

To achieve this each time, the use of a tripod is necessary. In the vast majority of cases, a tripod is quite often an integral part of the telescope package that is offered to consumers.

It is very important that the tripod be strong and robust. Any shake can seriously affect the proper viewing of objects. And remember this: the further away the object being observed is, the greater the magnification of any shakes induced.

You will see this warning again in this book, but it is important enough to repeat: The higher the magnification, the more pronounced any shake in the telescope will be. Additionally, any optical defect in the lenses will also be more pronounced.

If you have a telescope that is, say, two (2)-feet in length mounted on a tripod in windy conditions, the shake caused by the wind striking the telescope can be quite severe.

If you have a tripod with a center post, and you use it to elevate the telescope, the shaking can be even worse.

In this case, the telescope is supported by only one (1) post.

Avoid using this center post, especially if

there is even the slightest breeze about.

A sturdy tripod is essential for a telescope.

The bottom line with tripods is the stronger they are, the better.

Mounts

Now that we have covered tripods, we had better have a look at the two types of mounts that are generally available to the public.

The mount is the apparatus that connects the telescope to the tripod and allows you to adjust the direction at which you aim it.

In simple terms, the telescope needs to move in both a lateral direction—that is left to right—and a vertical direction—up and down.

Unfortunately, as with many things in life, there is more to it than meets the eye.

You see, the rotation of the earth causes objects to rise in the east and follow a <u>circular path across the sky</u>, in a westerly direction.

The popular saying, "Go west, young man," also applies to objects in the heavens!

This natural phenomenon causes any celestial object you are observing to keep moving out of sight, towards the west.

In fact, it only takes around 4 seconds for an object being viewed at high magnification by a telescope to move from the field of view of the eyepiece.

So constant adjustment becomes necessary.

With telescopes, the terms used for these movements are "right ascension" (RA), which covers lateral or side-to-side movements in a circular path across the sky, and "declination" (DEC), for position above or below the celestial equator.

These movements are expressed in hours, minutes and seconds for right ascension and degrees, minutes and seconds of arc for declination.

These are the road signs of astronomy, which enable you find stars, planets and other objects in the vastness of space.

You will read more on this in the chapter entitled, "Finding Your Way Around the Night Sky" (p. 61).

I mentioned earlier that there are two mounts that are commonly available for telescopes.

The first is called the *altazimuth mount,* which includes what are known as the Dobsonian mount, and the second is called the *equatorial mount.*

The Altazimuth Mount

The altazimuth mount (Alt-Az) is one that can move freely in two directions, horizontally (altitude) and vertically (azimuth).

Two controls allow these major movements, either horizontal or vertical, to position the telescope.

Normally, altazimuth mounts also have an additional control or two, called slow motion controls, to fine-tune the position of the telescope.

They are usually fairly rugged and light, and are ideal for terrestrial observations. But unfortunately, altazimuth mounts are not as good for astronomical observations as the equatorial mount, since a problem can arise when pointing the telescope directly overhead at the zenith.

In addition, unless they are computerized, they cannot track in the same manner as an equatorial mount.

The Altazimuth Mount
This mount allows lateral movement (left to right) (1) and vertical movement (up and down) (2). On this model, a control wheel allows fine adjustments on the lateral plane (3).

A Dobsonian mount, which has an altazimuth design.

The Dobsonian mount is a form of altaz-imuth mount that was developed in the United States by John Dobson.

Like the traditional altazimuth mount, it moves in two directions to allow positioning.

It is a simple design that is easy and cheap to manufacture and easy to use when combined with the Newtonian telescope, itself the cheapest and most simple in design.

It is also the simplest mount to set up, and therefore ideal for beginners.

A reflector telescope mounted on a single-fork mount.

A dual-fork altazimuth mount combined with a Schmidt-Cassegrain telescope.

Another type of altazimuth mount is the fork mount, which are very common on catadioptric telescopes.

There are two types of fork mounts, the single- and dual-, or twin-, fork mounts.

With correct use, you can point the fork arms towards either the North or South Celestial Pole, making it very convenient to use for astrophotography. But the fork mount needs to be computer driven to track properly.

Some astrophotographers will not use telescopes that have single-fork mounts, as they feel that the additional weight of a 35mm Single Lens Reflex camera body to the telescope would not be adequately sup-ported by one fork.

The fork mount is much lighter and more compact than the equatorial mount and as such is far simpler to set up and use.

Fork mounts can be converted into equatorial mounts with the addition of a wedge (see p. 33).

The Equatorial Mount

The equatorial mount was developed specifically for astronomical observations, and is also known as the German Equatorial Mount.

Earlier, we discussed the circular track of heavenly bodies due to the earth's rotation.

Due to this movement, objects that start out being dead center in the telescopes eyepiece can very quickly move from the field of view.

To counter this using an altazimuth mount, you need to constantly adjust both your telescope's horizontal and vertical positions either by hand or with the aid of a computer. This can be either very involved or expensive.

However, with the equatorial mount, the telescope is able to track a heavenly body in its circular path across the skies with only one adjustment of right ascension (RA), which is on the polar axis.

The polar axis of the telescope must be aligned to either the North or South Celestial Pole (see Using the Telescope).

The right ascension can then be adjusted by turning a gear by hand, or automatically, by using a motor drive.

The purpose of the motor drive is to allow you to follow an object as it traces its path across the night sky.

This simplifies the task of keeping the object you are observing squarely in the center of the eyepiece.

It is especially helpful in astrophotography, which naturally requires long exposure times.

As any photographer knows, if crisp, sharp images are what you are after, shooting a moving object during long exposures is not ideal.

The equatorial mount can be adjusted in four directions, as follows:

1   The polar axis. On the telescope, this is locked onto the North or South Celestial Pole, and is therefore parallel to the earth's axis.

This axis is used to track an object's elliptical path (RA) across the sky, or its right ascension. You adjust it by using the RA knob on the mount or by using a motor drive.

2   The declination (DEC) axis. On the telescope, it is at a right angle to the polar axis. Declination refers to the position of a celestial body, in degrees, either above or below the celestial equator (see Glossary).

3   The vertical adjustment (up and down). This is different to the declination.

4   The horizontal adjustment (side to side). This is controlled in different ways depending on the particular model of mount.

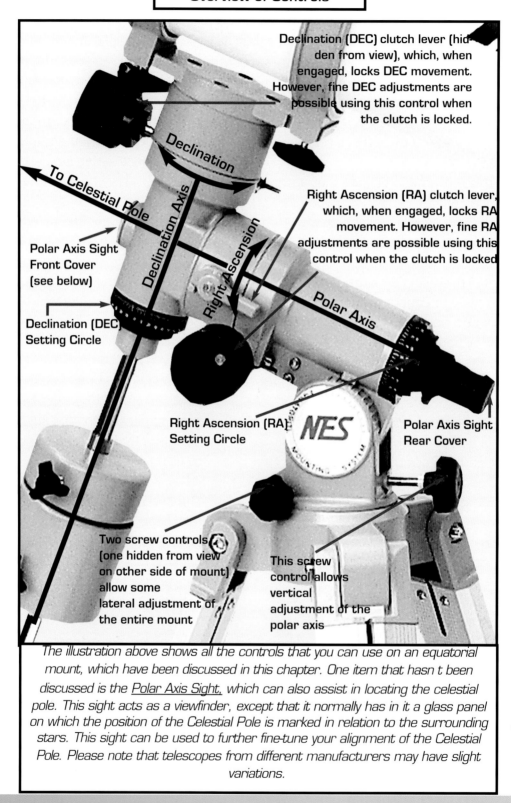

Declination (DEC) clutch lever (hidden from view), which, when engaged, locks DEC movement. However, fine DEC adjustments are possible using this control when the clutch is locked.

Declination

To Celestial Pole

Declination Axis

Right Ascension (RA) clutch lever, which, when engaged, locks RA movement. However, fine RA adjustments are possible using this control when the clutch is locked

Right Ascension

Polar Axis Sight Front Cover (see below)

Declination (DEC) Setting Circle

Polar Axis

Right Ascension (RA) Setting Circle

NES

Polar Axis Sight Rear Cover

Two screw controls, (one hidden from view on other side of mount) allow some lateral adjustment of the entire mount

This screw control allows vertical adjustment of the polar axis

The illustration above shows all the controls that you can use on an equatorial mount, which have been discussed in this chapter. One item that hasn t been discussed is the <u>Polar Axis Sight,</u> which can also assist in locating the celestial pole. This sight acts as a viewfinder, except that it normally has in it a glass panel on which the position of the Celestial Pole is marked in relation to the surrounding stars. This sight can be used to further fine-tune your alignment of the Celestial Pole. Please note that telescopes from different manufacturers may have slight variations.

The equatorial mount is easy to distinguish, as it has counterweights that help balance the weight of the telescope.

Eyepieces

This section deals with the eyepiece, which allows you to view the scene produced by the objective lens or primary mirror at the focal point.

Each eyepiece has an apparent and an actual field of view, which can be confusing, but I will try to explain this as simply as possible.

An *apparent field of view* refers to the apparent width of the view and this is calculated by the angle your eyeball must move from side to side to fully view the scene when using a particular eyepiece.

A selection of eyepieces

Eyepieces are the second lens in a refractor telescope and the only lens in a reflector telescope.

Eyepieces are normally described by their focal length and design style, for instance a 26mm Plössl. The 26mm is the focal length and Plössl is the design.

The focal lengths of eyepieces range from 2.5mm to 60mm plus.

This apparent angle of view varies with different eyepiece designs and can vary from between 30 to 80 degrees.

The important thing to remember here is that if an eyepiece has an apparent field of view of 50 degrees <u>that does not mean</u> that you will see a 50-degree arc of sky.

The amount of sky you see is the *actual field of view,* and this is calculated by dividing the apparent field of view by the magnification factor of the eyepiece.

To explain further, let's say you have a Plössl eyepiece in your telescope with an apparent field of view of 50 degrees. This same eyepiece, combined with your telescope, has a magnification factor of 80x.

If you divide the apparent field of view, which is 50 degrees by the magnification factor of 80x you get 0.625 degrees (50÷80 = 0.625).

Therefore, the actual field of view is 0.625 degrees, which is around two-thirds the width of a full moon.

Please note that the eyepiece does not magnify the object you are viewing. This task is performed by the telescope. Rather, it magnifies the image of the object you are viewing, created by either the objective lens or primary mirror.

This is a very important distinction and I cannot impress it upon you too much.

Salespeople pushing the magnification factor of the eyepiece have fooled too many people into buying inferior telescopes.

What is the point of magnifying an image with a high magnification eyepiece when that image is poorly presented due to the telescope's small aperture or inferior optics? All you end up with is a bigger view of a bad image.

There is another factor to consider here. When you increase the magnification of an image by the eyepiece, the amount of light that is available to produce the image is not increased, it stays the same.

So as your image increases in size, it decreases in brightness. The light that is available to produce the image is spread over a greater area.

In fact by doubling the image size, you reduce its brightness by a factor of 4 times (4x). Remember, the telescope's job is to gather light, and if it can't do this, then you won't see much no matter how much magnification power you use.

With objects in space, which are not bright to begin with, it is possible to lose the image entirely by magnifying it. You use one eyepiece and it's there, you use another and it disappears. This is not magic, it's physics.

Another factor to consider when it comes to eyepieces is *eye relief.*

Eye relief describes the distance between the outermost glass surface of an eyepiece and the lenses in your eyes.

If you wear glasses, this space must be big enough to accommodate them. Plössl eyepieces are among those that have good eye relief.

If you don't wear glasses, you have to remember not to get so close that your eyelashes come in contact with the eyepiece, thus dirtying it. Another risk of being too close to the eyepiece is that by coming in contact with the telescope, you can cause it to shake.

The white disk of light seen on this eyepiece is its exit pupil, when attached to my telescope. A different telescope could produce a different size exit pupil.

You should investigate the eye relief matter fully before selecting a particular eyepiece.

Lastly, you need to look at the exit pupil size of the eyepiece.

Exit pupil size is covered in the chapter on binoculars, but I will discuss the what's relevant to telescopes here.

Exit pupil refers to the size (diameter) of the light that comes out of an eyepiece.

For a telescope, it is calculated by the aperture size divided by the magnification factor of the eyepiece being used.

So, if your telescope has a 100mm aperture and your eyepiece has a magnification of 58x, then your exit pupil size is 1.72mm (100÷58 = 1.72mm).

Depending upon a number of factors—including your age and whether or not you smoke—your eyes pupils, at night, could be between 4–7mm in diameter.

If the exit pupil of the telescope's eyepiece is smaller than the pupil of your eyes, then your eyes can utilize every ray of light that passes through the telescope's optics.

If, however, the exit pupil of the eyepiece is larger than your pupils, some of the light passing through the telescope is lost and image brightness is reduced.

However, with telescope eyepieces, this is unlikely, since their exit pupils are generally not large.

Having said all that, I must stress that the use of a good quality eyepiece is essential for good viewing and many a telescope owner has had the performance of their equipment significantly improved by upgrading the eyepiece.

It is a matter of being informed of the various factors and selecting carefully. Each telescope needs an eyepiece and, generally, at least one is supplied with the telescope.

Eyepieces are categorized by their magnification ability. This is expressed in millimeters, for example, 10mm.

The degree of magnification increases with the reduction in millimeter size. An eyepiece designated as a 25mm model will not magnify the image as much as one designated 10mm.

To calculate the magnification factor of an eyepiece, simply divide the focal length of the telescope by the focal length of the eyepiece.

For instance, let's say your telescope has a focal length of 1000mm and you are using a 20mm eyepiecel.

Divide 1000mm by 20mm and you arrive at a figure of 50. That is the magnification factor of that particular eyepiece when used with that particular telescope. This is expressed as 50x.

If the same eyepiece (20mm) were used with a telescope with a focal length of 1500mm, then the magnification factor would be 75x (1500mm÷20mm = 75x).

Astronomers generally do not recommend magnification over 200x–250x, because the image quality can degrade at this level and upwards.

However, even this depends on numerous factors, including the condition of your eyes, the quality of your telescope's optics and atmospheric conditions at the time of viewing. It's a general rule and should be adapted for individual circumstances.

Most keen amateur astronomers will have around three eyepieces in their equipment bag.

This gives them three variable magnification levels—low, medium and high power.

For instance, to use with a telescope that has a focal length of 1500mm, they might have a 26mm (57x) for low power, a 17mm (88x) for medium power and finally a 9mm (167x) for high power.

There are many eyepieces on the market, with different optical designs (and costs).

In addition, they come in three different diameters, or barrel sizes.

These sizes are 0.965-inch (24.5mm), 1.25-inch (31.8mm) and 2-inch (50.8mm).

To be precise, there is a fourth which is 0.917-inch (23mm), but it is not that common, so I am ignoring it in this book.

Many entry-level telescopes come with a 0.965-inch eyepiece, but they can often be of dubious quality.

Simply by upgrading to a Kellner or Orthoscopics eyepiece, you can make a difference in the quality of image you see.

The more serious telescope manufacturers either use the 1.25-inch or 2-inch eyepiece, which are much better viewing platforms. The 2-inch varieties will, however, only fit on top-of-the-line models.

Eyepieces vary in terms of the number and types of lenses, or elements, they use. In the category of 1.25- to 2-inch eyepieces, they range from the Plössl design, which is a 4- element group, to the Nagler design, an 8-element group, and many others.

Plössl eyepieces are standard on many telescopes and are fine eyepieces. Talk to an expert when it comes to eyepieces to get a real feel for what's available and what they can do for your viewing quality. If possible always try, before you buy.

Go to a star party and look through other people's equipment and see how various eyepieces of different focal lengths and designs perform.

Better still, set up your own telescope and borrow (if possible) other eyepieces and try them in your equipment.

You'll soon get a feel for how different eyepieces perform and the results that you can expect when they are used with your equipment.

More importantly, you'll also save wasting your money on eyepieces that don't perform correctly for your equipment and viewing needs.

Barlow Lens

With telescopes, when you use eyepieces with a short focal length (and high magnification) the eye relief of the eyepiece is also shortened, which makes it less comfortable to use.

Simply put, the smaller the focal length of the eyepiece the closer the observer has to position him or herself to it.

This can be uncomfortable, especially if you wear glasses.

This occurs especially with eyepieces that have a focal length of between 4 to 8mm.

A 2x Barlow lens, which will double the magnification ability of any eyepiece.

The use of a *Barlow lens* can alleviate this problem.

A Barlow lens is, in simple terms, an extension tube that further increases the magnification factor of an eyepiece.

A typical Barlow lens has a magnification factor of 2x, meaning that it will increase the magnification by a factor of two.

They are available with multiplication factors from between 1.75 to 5. A few models have variable multiplication factors.

The Barlow lens is placed between the objective lens or primary mirror and the eyepiece.

For instance, a 12mm eyepiece attached to a 2x Barlow lens turns the view seen through the eyepiece from a 12mm to a 6mm view.

In other words, the image presented is magnified to twice the size with the addition of the Barlow lens.

The same eyepiece used with a 3x Barlow lens becomes a 4mm eyepiece.

Barlow lenses also double the number of effective eyepieces you have at your disposal, which, when considering the cost of good quality eyepieces, amounts to big savings.

Barlow lenses with magnification factors above 2x are sometimes criticized for poor quality. So be careful.

A dewcap attached to a refractor telescope.

Dewcaps

Have you ever noticed with camera lenses that some have a hood on the end that protrudes past the front of the lens?

These are called lens hoods and their job is to help shield the lens from light that is not coming directly from the subject being photographed.

Telescopes use something similar but, in their case, they are called dewcaps. Their prime function is slightly different, but still includes protection from stray light.

A telescope that is exposed to night conditions for a length of time can easily attract dew or frost.

The dewcap is designed to help protect against this and you will normally find that most refractor telescopes come with them as standard equipment.

With other designs you normally have to purchase a dewcap as an option, but do so, as it is worth it.

Filters

Sometimes you can improve the image you are viewing by the addition of filters.

There are filters to minimize light pollution called **LPRs** (Light Pollution Reduction).

These can reduce the effect of the glow in the sky that is produced by the ever-present light from cities and towns.

In addition, the night sky gives off its own light caused by the earth's atmosphere being excited by sunlight during the day.

At night, when the temperature drops, the energy stored by this activity during the day is released, producing small amounts of light, reducing a telescope's effectiveness.

LPRs are effective in combating this light source, as well. For this purpose, you need to specify that you need a *broadband LPR,* when you are making your purchase.

Narrowband LPRs perform a different function (see Nebula Filters, later in this section). Broadband LPR filters cover a wide section of the light spectrum, whereas narrowband LPR filters only affect a narrow section of the light spectrum.

The use of **colored filters** can accentuate certain colors while viewing.

For instance, a green filter will usually improve an object that is red (not the sun, however; see next section). A red filter will help with dark planetary markings, as seen on Mars.

The use of **gelatin filters** is also helpful for observing with a telescope.

For instance, the use of a UV- (ultra violet) transmitting gelatin filter called a Wratten 18a is helpful when viewing cloud formations on some planets.

Other useful gelatin filters include the following Wratten types: The Wratten 44a (blue), W58 (green), W15 (yellow), W25 (red) and finally a W23a (orange).

To gauge a filter's effect, just hold it up to the eyepiece and you will see how the filter affects the image being viewed.

Please remember that under no circumstances should you try to view the sun with any optical instrument whatsoever (including your eyes) without the use of specially-designed filters.

Solar Filters are a must when trying to view the sun.

<u>**Doing so can, and will, lead to severe, permanent damage to your eyesight and can also damage your equipment's optics.**</u>

Use only solar filters that cover the entire aperture of the telescope. With binoculars, both apertures must be entirely covered by solar filters.

Two types of filters can be used to view the sun.

The first is a *neutral density filter,* used to view or photograph the sun's surface.

The second is called a *Hydrogen-Alpha filter,* and is used to view or photograph the violent eruptions that occur in the sun's chromosphere, which is a layer of the sun's atmosphere.

There are a variety of Hydrogen-Alpha filters that are available, each having a different *bandpass*.

The bandpass of a filter describes how much light, on either side of the band, is transmitted and how much is reflected.

These filters are not inexpensive and the really high-quality, narrow-bandpass varieties, which give stunning images can cost as much as a computer-controlled Schmidt-Cassegrain telescope.

There are also filters called **moon filters** which are also used, when viewing the Moon, to reduce the bright light. This has the effect of increasing the amount of detail that can be discerned.

There are two types of moon filters. One is the *neutral density filter,* which reduces the amount of light by one factor, and the *polarizing filter,* which works in the same way as the sunglasses by shutting out light that is not aligned to the filter.

Moon filters also work well when viewing or photographing Jupiter and Venus.

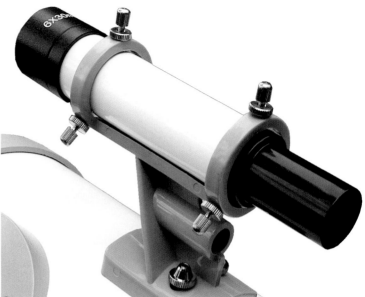

A finderscope is used to help find your target in the heavens, but must be aligned with the telescope to be effective.

Nebula filters are used to enhance the contrast between nebulas and the dark skies around them for better viewing.

There are two types of nebulae known as *emission* and *planetary nebulae.*

Emission nebulae occur where stars are born and planetary nebulae occur where stars have died.

They both emit light in discreet wavelengths, some of which are hard for the human eye to see (such as red light at low levels).

The more popular type of nebula filter is known as the Oxygen-III (O-III) filter, which enhances the bulk of the light given off by nebulae.

Nebula filters are what's known as *narrowband LPRs,* as previously noted. They darken the background glow of the sky sig-

nificantly, without affecting the view of the nebula.

Finderscopes

The finderscope is the small secondary telescope that attaches to the main telescope and is used to help you locate objects that you wish to view.

Many budget telescopes come with a 5 x 24 finderscope. This means the finderscope has a magnification factor of 5x and a 24mm aperture.

A finderscope of this size is fine for finding large objects, but can be less helpful when it comes to objects that are small and much fainter.

You could consider upgrading your finderscope to a 6 x 30 or 8 x 50 model to help find those elusive small objects in the night sky.

A basic computer software package that helps teach you about the heavens and also locate specific targets.

Software

There are many software packages out there that can increase your knowledge and enjoyment of the night skies.

One of the most useful program types are what's known as the planetarium packages.

Once set up on your computer and calibrated with the time, date and your location, they can then give you a view of what's showing in the skies that night.

They allow you to print star charts to take outside with you, which can be a real plus.

Some of the more advanced packages also allow you to control certain telescopes directly from the keyboard.

Wedges

To correctly align your telescope, you first point it either toward the North or South Celestial Pole, depending on where you are located (see Setting Up Your Telescope).

Schmidt-Cassegrain, Maksutov-Cassegrain and telescopes on equatorial mounts are all in this category, unless they are GPS models.

To perform this task, the first thing you need to know is the location of the Celestial Pole in question and position your telescope accordingly. Note that the North or South Celestial Pole is not the same as the North or South Magnetic Pole.

Secondly, you need to point the telescope to the specific position in the sky where the Celestial Pole is located.

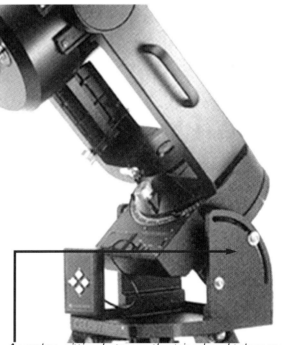

A wedge, sitting between the tripod and telescope and adjusted to the present latitude

To accomplish this, you need to know the latitude of your location.

The latitude of your location determines the altitude at which to angle your telescope.

If you don't know the latitude of your location, you can use the following website to find it. Log on to http://www.heavens-above.com/countries.asp.

This website will give you the answers and also provides a wide range of information for the astronomer.

Let's say you're located in Sydney, Australia. The latitude of Sydney is 33° 8830. That's 33 degrees, 88 minutes and 30 seconds.

So, to correctly align your telescope to the South Celestial Pole, the altitude should be set to 33° 8830.

This is where the wedge comes in.

The wedge is simply a device that either sits on the ground or, alternatively, between the tripod and the telescope mount.

It is made so that it produces an angle equivalent to your latitude, or in some cases, is adjustable so that it can adapt to various locations.

Aligning to the Celestial Pole restricts the telescope's ability to change altitude.

Using a wedge eliminates this loss of adjustment capability, and the finer the

degree of adjustment the better.

The wedge also makes the setting up of the telescope much quicker and easier.

When using motor drives on the Schmidt-Cassegrain and Maksutov-Cassegrain telescopes, there is another advantage to using a wedge.

That advantage is that your telescope uses less battery power when tracking, and the tracking itself is more precise.

A wedge transforms a fork mount, which is an altazimuth mount, into an equatorial mount.

Planisphere and Star Charts

A planisphere is a flat chart of the night sky that can be adjusted for time and date.

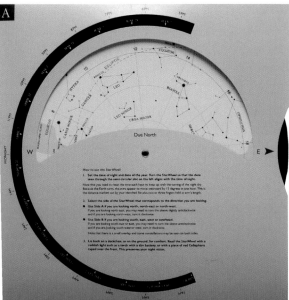

A planisphere is an excellent and economical tool for the amateur astronomer. Simply set the date and time using the wheel, and you will have laid out the stars visible at that time .

Therefore, you can use a planisphere at night to help you find objects in the night sky or just to familiarize yourself with the heavens.

You don't even need to have a telescope. A planisphere is a perfect accessory for viewing the night sky with your naked eyes or binoculars.

A star chart can be quite complex and detailed and, depending on the degree of complexity, can show objects that are only visible through the most powerful telescope. They are the road maps of the night sky.

You should learn to use a planisphere before graduating to star charts.

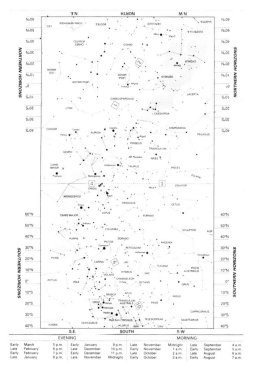

A page from the Bright Star Atlas showing star coordinates at particular times

Red Flashlight

Because astronomy demands you set up and use equipment at night, preferably in very dark conditions, you can have a problem.

That problem is simple: How are you supposed to see what you are doing when it is so dark?

A red flashlight is essential for astronomy.

Well, you can use a flashlight to see what you need to see, but that has the unfortunate side effect of compromising your night vision for a minimum of 15 minutes.

Some astronomers say it really takes up to a hour for your eyes to adjust after being exposed to bright light.

A special red flashlight can solve this problem for you.

As the human eye is least sensitive to red light, you can expect to regain full night vision more quickly when using these flashlights in comparison to the ordinary flashlight.

Good quality red flashlights use an LED (Light Emitting Diode) instead of lightbulbs and will have an adjustable brightness range that suits the various tasks that are required in astronomy.

Collimation Eyepiece

One other accessory that is very handy, perhaps essential, to owners of Newtonian and Schmidt-Cassegrain telescopes, is a collimation eyepiece.

This piece of equipment is covered in some detail in a later chapter (see p. 48).

There are many other accessories available to the amateur. Some are referred to in other sections of this book and others have been left out.

What is covered here is enough to get you well on your way with your new pastime.

The remainder you will come across soon enough.

A collimation eyepiece, a very handy tool for simplifying the collimation of reflector and Schmidt-Cassegrain telescopes.

Setting Up Your Telescope

So you've read the preceding information on telescopes and made your purchase.

Congratulations! You are about to enter a world filled with wonder and beauty.

But before you do, you need to set up your telescope and work out what everything does.

So, let's go through the exercise of setting up a refractor, reflector and Schmidt-Cassegrain telescope just in case the manufacturer's instructions are either confusing, or worse still, nonexistent.

Please remember that every telescope is different. This chapter is a guide, but you may need supplemental information to suit your individual equipment.

We'll start with a simple refractor telescope with an altazimuth mount.

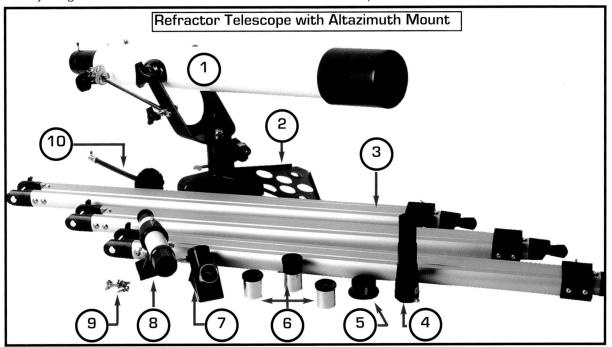

Refractor Telescope with Altazimuth Mount

Parts Identification List

1	Telescope tube with altazimuth mount and dewcap already fitted.	**5**	Adjustment collar for star diagonal.
2	Accessory tray (fits between the tripod legs).	**6**	Eyepieces (3).
		7	Star diagonal.
3	Tripod legs.	**8**	Finderscope.
4	Barlow lens.	**9**	Screws and wingnuts for securing accessory tray to tripod legs.
		10	Adjustment control.

Refractor Telescope with Altazimuth Mount

Even though this telescope and mount have a simple design and are fairly easy to set up, I opened the packaging and found that there were some 23 components that needed to be put together.

Assembling a telescope is fairly simple, as long as you remember that the main parts are as follows:

1. The tripod, on which the mount and the telescope sit.

2. The mount, which sits between the tripod and the telescope.

3. The telescope, which sits on top of the mount.

The first thing to do is to assemble the tripod, if that is required. Some tripods come already assembled.

This is not the case with the model that is illustrated here.

On this model, the tripod legs fit directly onto the altazimuth mount and are attached by using the wing nuts at the top of each tripod leg.

To afford extra stability, the accessory tray fits between the tripod legs and is secured by nuts, washers and bolts to each tripod leg.

As the telescope is already attached to the altazimuth mount on this particular model, the three main pieces of the unit are now assembled. On other models, though, you may have to fit the telescope to the altazimuth mount.

The finderscope fits into two retaining collars attached to the main telescope and is held by adjustable screws.

A finderscope needs to be correctly aligned (see Telescope Adjustments and Maintenance).

The next thing to do is to put the eye-piece and star diagonal in their place.

The star diagonal goes into the eyepiece aperture at the back of the telescope and the eyepiece goes into the star diagonal.

Your refractor telescope with altazimuth mount should look something like this when assembled.

Small screws on the telescope and star diagonal retain both.

With this particular telescope, there is an adjustment collar that sits between the star diagonal and the telescope's eyepiece aperture.

This can be removed, if you wish to use an eyepiece with a greater barrel diameter.

Lastly, there are two adjustment control knobs, one for RA and one for DEC. They are attched to rigid steel shafts that are part of the mount.

The Barlow lens and extra eyepieces supplied with the telescope, and any other accessories you might purchase, can be stored in the accessory tray for convenience.

Now that you have assembled your telescope, it's time to align the finderscope. This is covered in the next chapter.

Remember it's a good idea to check all the screws and wing nuts regularly, as they can loosen and make the telescope less stable.

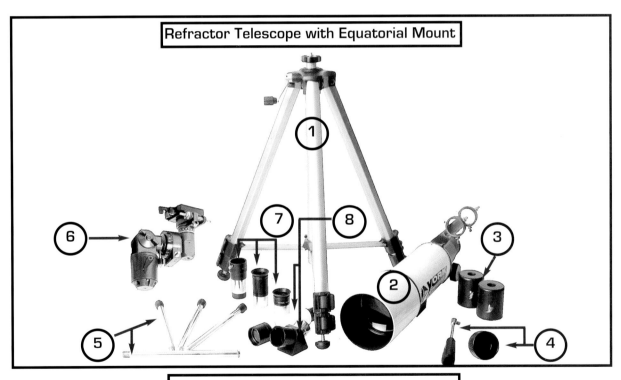

Refractor Telescope with Equatorial Mount

Parts Identification List

1	Tripod (pre-assembled).	5	Mini-tripod legs (3) with rubber end caps and counterweight shaft.
2	Telescope with dewcap fitted.		
3	Counterweights.	6	Equatorial mount.
4	Adjustment controls (2) (DEC and RA).	7	Eyepieces (3).
		8	Finderscope and star diagonal.

Refractor Telescope with Equatorial Mount (or German Equatorial Mount)

Although this telescope is more complex than the previous model, it is actually a bit easier to assemble, as some of the more important parts are pre-assembled, including the tripod and the mount.

So, first set up the tripod and then attach the equatorial mount.

In the photo (previous page), look for four lengths of silver, polished metal (item number 5), three of which have a rubber cap on one end.

These three are legs that can screw directly into the base of the equatorial mount, if you don't wish to use the tripod.

You may have a convenient, flat surface, like a table, that you can use. Not all models will come with this feature.

The fourth is a bar that also screws into the base of the equatorial mount to hold the counterweights.

Screw this bar into the mount. Then undo the screw at the base and slide the counterweights up the bar. Replace the screw at the base.

Be careful that the whole thing doesn't tip over, those counterweights are heavy (see Balancing Your Telescope, in the next chapter).

Now attach the telescope itself to the mount. With this model it's simply a matter of placing the mounting plate on the bottom of the telescope into the quick shoe on the mount and locking it in place with the lever.

Now attach the finderscope, star-diagonal, eyepiece and adjustment wheels in the same manner as the previously discussed telescope and you are done.

There are two adjustment controls: the one that adjusts RA should point towards the eyepiece at the back of the telescope.

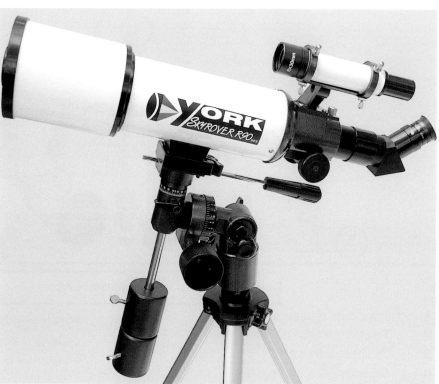

A refractor telescope with equatorial mount when assembled.

The one that adjusts DEC should extend from the right side of the mount, from the viewing position.

As with the previous model, you need to align the finderscope (next chapter) and also check the adjustment of any locking screws or nuts on the mount itself.

You should be able to lock the mount tightly in any position and alter it smoothly and easily with the adjustment wheels.

Reflector Telescope with Equatorial Mount

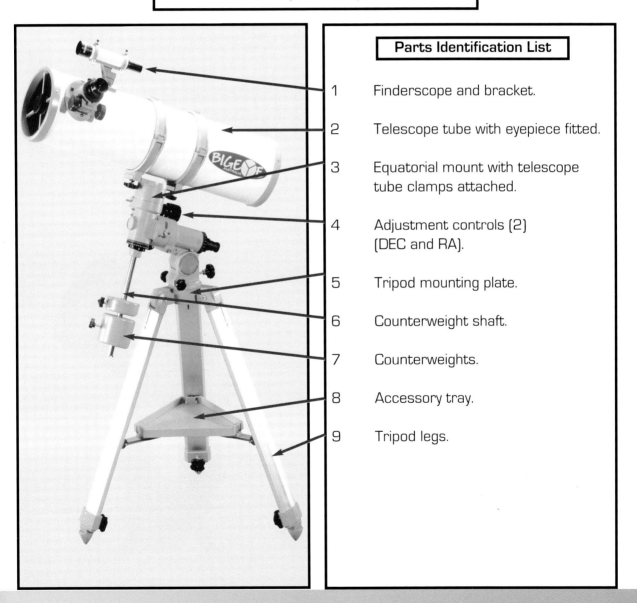

Parts Identification List

1   Finderscope and bracket.

2   Telescope tube with eyepiece fitted.

3   Equatorial mount with telescope tube clamps attached.

4   Adjustment controls (2) (DEC and RA).

5   Tripod mounting plate.

6   Counterweight shaft.

7   Counterweights.

8   Accessory tray.

9   Tripod legs.

Reflector Telescopes

To assemble a reflector telescope, follow the same general guidelines I gave for a refractor telescope with equatorial mount.

Using the model illustrated on the previous page, the sequence is as follows:

1   Locate the tripod mounting plate and secure the tripod legs to it.

2   Attach the accessory tray to the tripod legs.

3   Attach the equatorial mount to the tripod mounting plate You'll most likely find that the equatorial mount attaches in only one way.

4   Attach the counterweight shaft to the equatorial mount by screwing it in. Once done, slide the counterweights on and lock them in place. There will most likely be a removable lock on the bottom of the counterweight shaft that has to be removed before sliding on the counterweights. Make sure you replace this locking device.

5   Attach the telescope tube to the equatorial mount using the tube clamps.

6   Attach the finderscope bracket to the telescope tube and then attach the finderscope to the bracket. You are now all finished except for the alignment of the finderscope.

Schmidt-Cassegrain Telescopes with Fork Mount

This is without a doubt the easiest to set up of all the telescopes covered in this book.

Basically, there are only two major parts to put together—the telescope and its fork mount plus the tripod.

These will normally bolt together and once this is accomplished, all you need to do is add the finderscope, star diagonal, visual back and eyepiece.

The visual back screws into the back plate of the telescope, and the star diagonal and eyepiece attach to it.

Like the other telescopes, the finderscope needs to be aligned, which is covered in the next chapter.

A Schmidt-Cassegrain telescope on a dual-fork mount.

Telescope Adjustments and Maintenance

Now that your telescope is assembled, there are a few things to do so that it performs as it should.

Aligning the Finderscope

The job of the finder-scope is to assist you in finding the object you wish to view as easily as possible.

The finderscope has a wider field of view than does your telescope (by 5° degrees), so you can scan more of the heavens.

Ideally, once you have found the object you wish to view, and centered it in the finderscope, the object will also be centered in the telescope's eyepiece.

However, as we all know, it's not an ideal world and the chances are that some adjustments will need to be made.

This is known as aligning the finderscope.

By following the instructions below, you will be able to align the finderscope easily.

Set up the telescope outdoors in an area where you have an unobstructed view for at least one mile (1–2 kilometers) **The further away the better the accuracy.**

Pick a distant object such as a power line tower or any other distinctive object that you can readily identify and not confuse with any other object.

I cannot stress this enough: Make sure it is a distinctive object and easy to identify.

A finderscope on a refractor telescope Finderscopes need to be aligned to be effective.

Using the telescope, aim and focus on this object, making sure that it is dead center in the eyepiece.

Start with a low power eyepiece. Then swap it for a high power eyepiece to ensure the image is dead center.

Lock the telescopes adjustments so that it cannot move.

Check again to ensure your object is still dead centre in the eyepiece. If not, unlock the telescope and readjust until it is, then lock it up. Check the eyepiece again.

Now move to the finderscope. Normally it will have either three or six adjusting screws so that it can be adjusted.

These retaining screws and nuts must be loosened before you can align the finderscope.

These retaining screws most likely will also have locking nuts.

Loosen three screws and lock the nuts, so that the finderscope can move.

Look through the finderscope. By moving it around, try to center your object in the finderscope's crosshairs.

Remember, the finderscope is a small refractor telescope and as such your image will be upside-down and mirror-reversed.

If you cannot quite get there, see if you can give the three retaining screws some more slack by loosening them a tad more. In extreme cases, you may need to slacken the other three retaining screws as well.

Once the object is in both the finderscope's and telescope's sights, gently tighten the back retaining screws, so that the finderscope is held firmly. Check one more time that the view in the finderscope is still identical to the view in the telescope.

Now check the front retaining screws and tighten them, if necessary.

Do this while looking through the finderscope to ensure the view does not shift.

Check the finderscope's view again and compare it to the view through the telescopes eyepiece. If it is still identical, tighten the locking nuts at the back, and you're finished.

Note: The further away the object is, the more accurate the alignment of the finderscope will be.

Do not choose a star or planet to do this function, as it is constantly moving, thus making the task of aligning the finderscope nearly impossible..

Balancing Your Equatorial Mount

A telescope (any type) mounted on an equatorial mount needs to be balanced both on its right ascension (RA) and declination (DEC) axes, so that the instrument will function correctly and move smoothly.

Balance the declination axis first and then the right ascension axis.

This should be checked regularly because with use the equatorial mount can wear out, making shifts in position more likely.

However, balancing the RA and DEC axes is a simple task and can be performed quickly. Just follow these steps in the order shown:

Declination (DEC) Axis Balance

1   Set the telescope up on a firm surface, using adjustments of the tripod legs and a carpenter's level, or spirit level, to ensure it is perfectly level.

2   Release the right ascension axis clutch and move the telescope tube until both it and the counterweights are parallel to the ground. Now re-apply the right ascension axis clutch.

3   Hold on to the telescope tube and release the clutch knob for the declination axis. Let go of the tube, but be ready to grab it if it rotates towards the ground.

4   If the telescope tube does rotate towards the ground (either the front or back), then the declination axis is out of balance and must be adjusted.

5   If the front of the telescope rotates towards the ground, loosen the telescope tube clamps and slide the telescope tube towards the rear.

If the back of the telescope tube rotates towards the ground, move the tube towards the front.

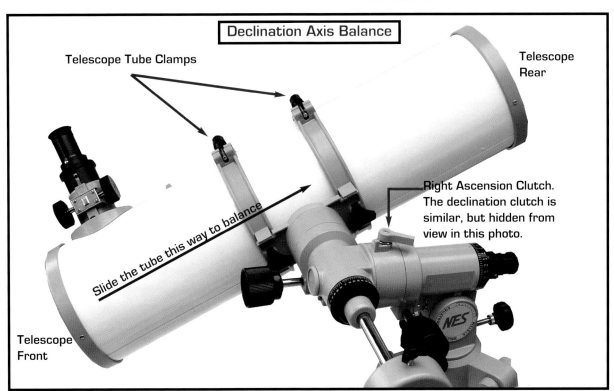

Declination Axis Balance

Telescope Tube Clamps

Telescope Rear

Slide the tube this way to balance

Right Ascension Clutch. The declination clutch is similar, but hidden from view in this photo.

Telescope Front

In this example, the front of the telescope has rotated towards the ground. The telescope tube needs to be moved towards the rear by releasing the telescope tube clamps, adjusting the tube and tightening the clamps again. Move the tube in the opposite direction, if the rear rotates towards the ground.

6. Repeat the procedure (steps 3 & 5) until the telescope tube holds its position and does not rotate.

7 Once the declination axis is in balance, re-apply its clutch knob.

Right Ascension (RA) Axis Balance

1 Ensure that the telescope is standing on firm ground and is perfectly level by use of tripod leg adjustments and a spirit level. Also, the declination axis clutch should be engaged.

2 Release the right ascension axis clutch and rotate the telescope until the tube and counterweights are parallel to the ground. Hang on to the telescope tube.

3 Release the telescope tube, but be prepared to grab it if it or the counterweights rotate towards the ground.

4 If the telescope tube rises, the counterweights are too heavy and need to be adjusted upwards towards the telescope tube. The opposite

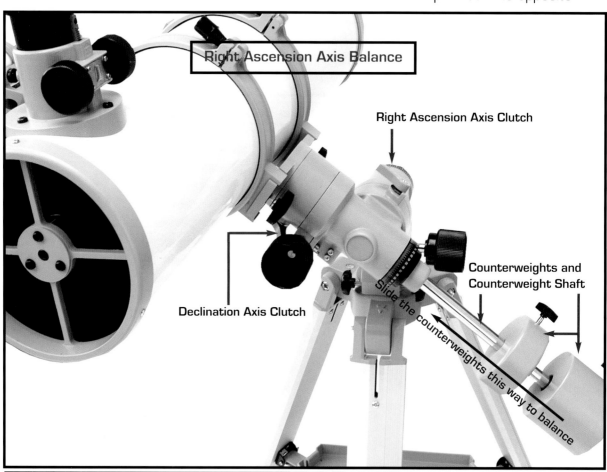

Right Ascension Axis Balance

Right Ascension Axis Clutch

Counterweights and Counterweight Shaft

Slide the counterweights this way to balance

Declination Axis Clutch

In this example, the counterweights have rotated towards the ground and thus need to be moved up the counterweight shaft to balance the right ascension axis. If the telescope tube had dropped towards the ground, the counterweights would need to be moved up the shaft in the opposite direction.

procedure is employed if the telescope tube rotates towards the ground.

5   Once you achieve balance, you are finished and can return the telescope to its normal position.

Note: The addition of accessories such as camera, solar filter, and other items can change your telescope's balance, and you will have to re-adjust it accordingly.

Balancing Your Dobsonian Telescope

Balancing a Dobsonian telescope is simple, but important, as the Dobsonian relies on friction to hold its position.

Too much friction and the telescope is difficult to move, making fine position adjustments difficult during use.

Too little friction and the telescope will not hold its position correctly.

Dobsonian telescopes have a number of different balancing systems, but you should get the idea from the following:

1   First set up the telescope with all the accessories you want to use Make sure it is level on the ground. Use a spirit level to check this.

2   Adjust the telescope tube so that it is pointing at 45 degrees.

3   Let the tube go. If it is badly out of balance, it will move on it own accord,

A Dobsonian telescope at 45 degrees.

either up or down.

4   Don't let the front of the tube hit the ground. If the tube does not move of its own accord, give it a gentle push, first up and then down.

5   An out of balance Dobsonian will feel harder to push in one direction than another.

6   If the telescope sinks towards the ground or feels easier to push down rather than up, it is front-heavy. The opposite indicates it is back-heavy.

7   If the telescope is front-heavy, it needs to be adjusted towards the back (if possible) or weights need to be added, depending upon the make. Here, you need to check the literature from the manufacturer. If it is unclear, check with the retailer who sold you the telescope.

8   If the telescope is back-heavy, it should be adjusted towards the front or weights need to be either added to the front or removed from the back.

9   Once adjustments are done, repeat steps 2–6 to check balance. Fine-tune if necessary by repeating steps 7–8.

Note: Like the telescopes discussed earlier, wear and tear, or the addition of accessories, can throw off the telescope's balance. Check this regularly, especially if adding, say, a camera body or other weighty accessory.

Collimation

Collimation is the precise alignment of the optics in your telescope, so that the image presented at the focal plane is as perfect as it can be.

The optics that we are referring to are the primary and secondary mirrors on a Newtonian and the secondary mirror **only** on a Schmidt-Cassegrain telescope.

If you have a refractor telescope, the optics are factory aligned and cannot be adjusted by the owner.

If your refractor telescope's optic alignment is out of kilter, send it back to the manufacturer for correction.

Maksutov-Cassegrain telescopes are noted for their optics stability and should not require collimation.

Collimation is necessary when the optics are thrown out of alignment, often caused by incorrect handling during transport.

Many people find collimation difficult, but if you take your time, it should not be a major chore.

Some Points to Remember

1   Each mirror has three collimation screws, and the primary mirror on a Newtonian may also have locking nuts.

2   It can be helpful if the various adjustment screws are identified, perhaps with a different colored sticker.

3   Always make adjustments, where possible, looking down the eyepiece barrel so that you can immediately see the effect of each adjustment.

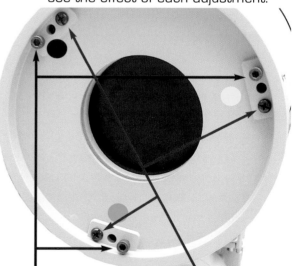

The three locking nuts and collimation screws for the primary mirror of a Newtonian telescope. The collimation screws for the secondary mirror are at the front and similar to the Schmidt-Cassegrain illustration on the next page.

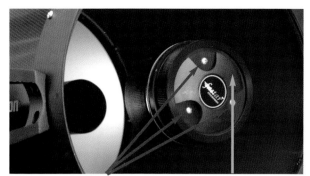

The three collimation screws for the secondary mirror on a Schmidt-Cassegrain telescope. The cover rotates to hide the screws when not being adjusted.

You will be surprised how far your arms stretch!

4 Only make adjustments in small increments, say a quarter of a turn clockwise or counterclockwise at a time.

5 You should only adjust a maximum of two screws at any one time.

6 It can also be helpful to record the adjustments you make as you go.

Collimation Tools

There are a number of tools on the market to assist the telescope owner with collimation, including a laser guide and a collimation eyepiece.

The collimation eyepiece is the one I have chosen to explain in this book, but the others will work just fine.

A collimation eyepiece is a combination of a "sight tube" and a "cheshire eyepiece".

The sight tube section has crosshairs at

one end and a sight hole at the other. These have two purposes.

The first is to accurately define the center of the optical axis and the second is to ensure the secondary mirror (on Newtonians) is centered directly under the focus tube.

The cheshire eyepiece has a polished element angled at 45 degrees. Its purpose is to reflect light from the hole in the side of the tube down into the optical path.

This highly polished flat element projects a bright ring of light, which provides a target for aligning the primary mirror.

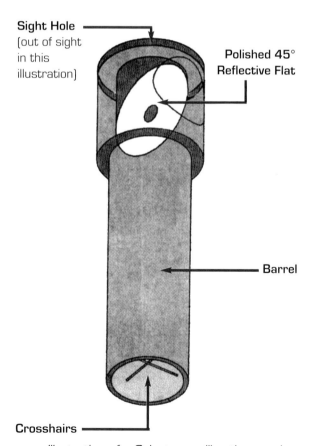

Illustration of a Celestron collimation eyepiece

Collimating a Newtonian Telescope Using a Collimation Eyepiece

When collimating a telescope, please ensure that the telescope or the cutout piece of the collimation eyepiece **is not pointing at the sun, otherwise you may incur serious eye damage**.

You can collimate at night by shining a flashlight through the cutout on the collimation eyepiece.

To collimate a Newtonian telescope using a collimation eyepiece, follow these steps:

1   Remove the eyepiece or any other accessory from the focus tube, or focuser.

2   Adjust the focuser, so that it is pushed inwards as far as it will go.

3   Insert the collimation eyepiece into the focuser far enough so that the bottom edge of its barrel is slightly larger than the outer edge of the secondary mirror, when viewed through the sight hole.

4   Rotate the collimation eyepiece, so that the cutout on the side of the barrel is pointed towards an external light source (**not the Sun**).

5   Look through the sight hole and determine if the secondary mirror is positioned directly under the focuser.

The center of the secondary mirrors should be directly beneath the intersection of the crosshairs of the collimating eyepiece.

If it isn't, you need to adjust the secondary mirror so that it is centered on the crosshairs.

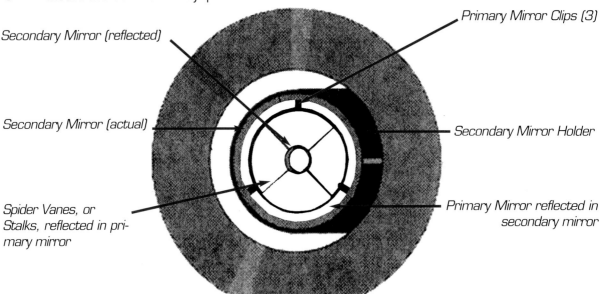

Secondary Mirror (reflected)

Secondary Mirror (actual)

Spider Vanes, or Stalks, reflected in primary mirror

Primary Mirror Clips (3)

Secondary Mirror Holder

Primary Mirror reflected in secondary mirror

When you look down the focuser of a Newtonian telescope with the eyepiece removed, this is what you should see, if it is correctly collimated.

This can be done by rotating the threaded rod that secures the secondary mirror holder. You can also slide the threaded rod up and down through the holder to adjust the other axis.

Stalk-style secondary mirror holders have a center bolt that can be loosened to allow the holder to be rotated.

6 Now you need to turn your attention to the tilt of the secondary mirror, so that the entire reflection of the primary mirror is visible in the secondary mirror.

Simply adjust any two of the collimation screws for the secondary mirror.

Make small adjustments (no more than a quarter of a turn at a time) to

one screw while looking through the sight hole to gauge the effect your adjustment has made.

Adjust the collimation screws until the reflection of the primary mirror is centered on the collimation eyepiece's crosshairs.

7 Lastly, you need to collimate the primary mirror so that the reflection of the secondary mirror is perfectly centered in the primary mirror. You should study the illustration on the next page.

Again, slowly adjust any two of the collimation screws (loosen the locking nuts first) for the primary mirror, while observing what's happening through the sight hole.

If you cannot look through the sight hole and adjust at the same time,

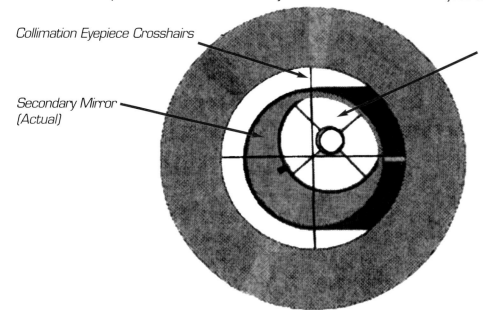

Collimation Eyepiece Crosshairs

Primary Mirror reflected in secondary mirror

Secondary Mirror (Actual)

Once you have adjusted the position of the secondary mirror, you need to adjust its tilt so that you can see the entire primary mirror reflected in the secondary mirror.

Remove the eyepiece and star diagonal from the back of the telescope and insert the collimation eyepiece.

have someone do the adjusting for you, while you observe.

Remember to instruct them to make small adjustments and to remember what the adjustments were, in case they need to be reversed.

8   When you are satisfied, tighten the locking nuts again.

Collimating a Schmidt-Cassegrain Telescope Using a Collimation Eyepiece

With a Schmidt-Cassegrain telescope,

you only need to collimate the secondary mirror.

Just follow the steps listed below:

1   Remove the star diagonal and eyepiece from the back of the telescope and insert the collimation eyepiece.

2   Looking through the collimation eyepiece's sight hole, you will see that the shadow of the secondary mirror will appear as a dark circle near the middle of the field of view.

3   Using two of the three collimation screws, adjust the secondary mirror until its reflection is centered on the crosshairs in the eyepiece.

Note: Remember the collimation screws will be found on the front of the

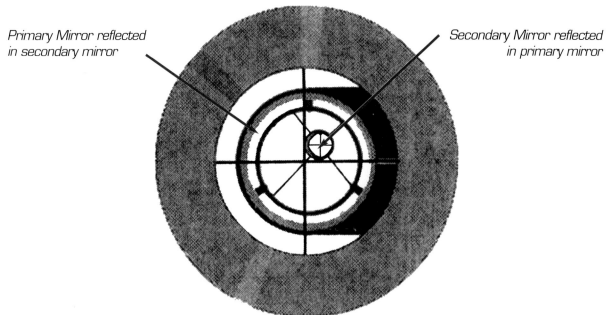

Primary Mirror reflected in secondary mirror

Secondary Mirror reflected in primary mirror

Finally, you need to adjust the primary mirror, so that the reflection of the secondary mirror is centered on the reflection of the primary mirror.

corrector lens. Only adjust a maximum of two at any time. Also, remember to make small adjustments and to watch the effect of each adjustment.

At the center of the corrector lens you may find another screw. Do not loosen this screw, as it may cause the secondary mirror housing to detach from the mount.

Collimation: Final Adjustments for Newtonian and Schmidt-Cassegrain Telescopes

To fine-tune either your Newtonian or Schmidt-Cassegrain telescope's collimation, follow these easy steps:

1   At night, find and focus on the brightest star in the sky at the time, preferably one that is not twinkling.

2   Make sure that the air temperature in the telescope tube has had time to equalize with the outside air temperature (allow about 45 min.).

For Newtonians, use steps 3 and 4 only. For Schmidt–Cassegrains, use 5–8 only.

3   Using a high power eyepiece, focus on the star you have selected and then defocus. You should have concentric circles surrounding the star. If they are not concentric, you need to tweak the collimation and make some minor adjustments.

4   While looking through the eyepiece, make very minor adjustments to the collimation screws (a maximum of 2), until the light circles are concentric.

5   Use a high power eyepiece and focus again on the selected star.

6   Now take the eyepiece out of focus until the shadow of the secondary mirror becomes a black obstruction in the image of the star.

7   If this black obstruction is not precisely at the center of the light circle, you need to make minor adjustments.

8   If correction is needed, do so with very fine adjustments of the secondary mirror's collimation screws (maximum of two), while looking through the eyepiece.

Telescope Care and Maintenance

Apart from collimation, there are a number of things that need to be done to keep your telescope in tip-top condition.

Keep your fingers and any other part of your body off the lenses and mirrors at all times.

If any part of your body comes into contact with the lenses or mirrors, it will leave oils on it which can eat into the optical coatings. Stay back!

Protect the telescope tube from shock, especially when moving it around.

Always store the telescope in a dry, cool environment with as little humidity as possible.

Use silica gel sachets when storing the telescope. These will help to absorb any moisture. You can purchase these from camera stores.

Keep the telescope away from extreme heat at all times.

Always use the dust (lens) caps when storing the telescope.

However, do no put the lens caps on the telescope until you are sure there is no moisture on the instrument, otherwise fungus may grow and ruin your optics.

If the optics do get dirty or dusty, only use lens cleaning solution to clean them, along with lens tissues. Do not use handkerchiefs or normal household tissues.

Lens cleaning solution and lens tissues, which, along with lens cloth (below), are useful for telescope maintenance.

A lens tissue is the only thing that should ever touch your optics!

A puff of air from a blower brush or a pressurized can of air can remove dust.

Both are available from camera stores.

If you use pressurized air, make sure you read the instructions carefully, otherwise you may release propellant, which is not good for your optics.

When using lens cleaning solution, do not pour it directly on to the lens.

Rather, pour a little onto a lens cleaning tissue and apply it in a straight line, starting from the center and working towards the edge.

Then use a fresh dry tissue in the same manner to wipe off the dirt and lens cleaner.

If you have to do this twice, use a new tissue. Never use the same tissue twice, unless the thought of scratches on your optics appeals.

Don't worry too much about dust on the primary mirror of a Newtonian. It will not affect the telescope's performance unless it becomes excessive.

The telescope tube can be cleaned with a soft cloth. If it is really dirty, use a very mild detergent to remove marks.

Collimation aside, don't fiddle with any other screws or bolts on the telescope, especially those holding mirrors in place.

Using the Telescope

Okay. You have given yourself a basic education on how to find your way around the skies.

You have also bought yourself a telescope, set it up and done everything necessary to ensure it will perform properly.

Now it's time to use it! Right?

Yes, but before you do, you need to set it up at the location where you plan to use it, be it your backyard or elsewhere.

Here are a couple of things to remember about the telescope, at this point.

Remember to transport it very carefully, if necessary, so that it suffers no bumps.

Choose your location carefully. The less ambient light in the skies the better, so, if possible, get as far away from cities and towns as possible.

At the same time, try to choose a location with as few obstructions—such as trees, mountains and power towers—as possible.

When you are ready to set up the telescope, you need to ensure that the air temperature inside the tube is the same as the surrounding air.

If it has been kept in a different temperature prior to use, give it time (about 45 minutes) to equalize the temperature.

When you set up the tripod, make sure it is on very firm ground and isn't gradually sinking.

Setting Up Altazimuth-Mounted Telescopes

Altazimuth-mounted telescopes, including Dobsonians and Schmidt-Cassegrain or Maksutov-Cassegrains (without a wedge), are a cinch to get ready for use.

Simply set up the tripod or base and then make sure it is absolutely level by using a spirit level and making minor adjustments to one or more of the tripod's legs.

Then attach the mount and telescope and you are ready to go. Happy stargazing!

Telescopes Fitted with an Equatorial Mount and Schmidt-Cassegrain or Maksutov-Cassegrains on a Fork-Mount with a Wedge (and No GPS).

Before they will perform properly, these telescopes have to be aligned to the poles. The idea is that, once this is accomplished, you can follow a star or some other object across the sky simply by turning the axis (by hand or motor drive) at the same rate as the rotation of the earth, but in the opposite direction.

For general viewing, the alignment can be off by plus or minus 5 degrees. But in astrophotography, which involves long exposure times, you are only allowed an error in accuracy of 0.2 degrees, beyond which you the photographed image will be blurred.

In the Northern Hemisphere, you need to align the telescope to the North Celestial Pole, and in the Southern Hemisphere, to the South Celestial Pole. But where are they?

This compass will help you find North or South but will not be much assistance with the North and South Celestial Poles.

Good question, but before we go into it, I should remind you where they are not—at the North or South Magnetic Poles.

North Celestial Pole (NCP)

The North Celestial Pole is the point in the sky around which all the stars seen in the Northern Hemisphere rotate.

It is not too hard to locate, because it is near Polaris (The North Star).

Aim the telescope at Polaris and match the angle of the telescope's polar axis to the latitude of your location. For example, if you are standing at 40 degrees latitude, the angle if you are in New York, the position of Polaris will be 40 degrees above the northern horizon.

You are now close enough to the Celestial North Pole for general viewing.

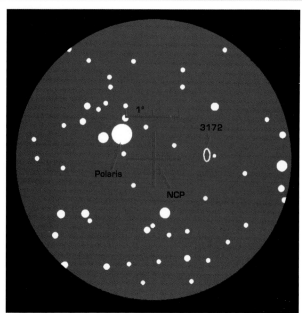

The North Celestial Pole (NCP) is just nine-tenths of a degree from the North Star (Polaris).

Although Polaris is not dead center at the Celestial North Pole, it is only nine-tenths of a degree away. [To refine this even further, see Fine-Tuning Polar Alignment.]

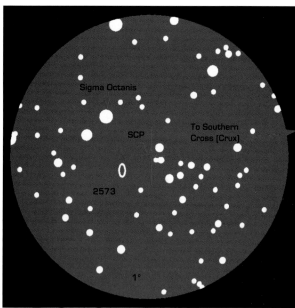

The South Celestial Pole (SCP) is harder to find than the North Celestial Pole. However, using the Southern Cross as a guide can get you pretty close (see later in this chapter)

South Celestial Pole (SCP)

The South Celestial Pole is the point in the sky around which all the stars seen from the Southern Hemisphere rotate.

People living south of the equator are not as lucky as their brothers and sisters in the Northern Hemisphere for, alas, they have no equivalent of Polaris.

The nearest major body is Sigma Octantis which is around 1.4 degrees away and not as obvious as Polaris.

In order to locate the Celestial Pole for your hemisphere, you first need to point

An ephemeris shows the position of objects in the solar system.

(align) your telescope either toward True North (in the Northern Hemisphere) or True South (in the Southern Hemisphere).

Once you accomplish this, you then need to elevate the telescope tube to an angle

that matches the latitude of your location. For instance, New York has a latitude of 40 degrees, so the telescope tube should be elevated to 40 degrees to locate the Celestial Pole.

To find True South or True North, first get hold of an ephemeris. This is an annual periodical that shows the position of the objects in our solar system throughout the year.

What it will also show is the precise time when the sun rises and sets every day throughout the year.

It will normally show two times: one for sunrise (BEG and RISE) and one for sunset (SET and END).

BEG indicates when the sun first breaks over the eastern horizon. RISE indicates when the sun has fully risen.

SET indicates when the sun starts to dip below the western horizon, and END marks the time when it has fully sunk below the horizon.

Pick a day when you wish to conduct this exercise. It must be sunny, with no cloud cover or strong wind.

Consult the ephemeris for the day picked and get the times for either BEG and END or RISE and SET.

Let's say you pick RISE and SET. You now need to subtract the time specified for RISE from the time specified for SET.

This will give you the number of hours and minutes between sunrise and sunset.

Divide the result of your subtraction by 2. This will give you the exact time of the midpoint of the Suns travels during that particular day.

At the position where you plan to set up your telescope, set up a structure of some kind with a horizontal arm about 3 or more feet (1 meter plus) from the ground.

From the horizontal arm, suspend a plumb bob, or plumb line, which is a length of line with a weight at the end that uses gravity to establish what is exactly vertical.

Make sure it does not touch the ground. The line should be reasonably thick, so that it will throw a perceptible shadow.

Mark the shadow of the plumb bob at the exact time the sun reaches its midpoint. The plumb bob should not be moving when you record its shadow.

The record you have just made of this shadow point in the direction of the True North or True South.

It will enable you to align your telescope to the Celestial Pole within about 2 degrees of accuracy, once you have adjusted the angle of the polar axis to the latitude of your location (see Fine-Tuning Polar Alignment).

Another Method of Finding the South Celestial Pole

If you look at the diagram of the Southern Cross (Crux) on this page, you will see that I have projected an imaginary line through the two stars that form the horizontal part of the cross and projected the line to the left towards the South Celestial Pole.

The bottom star of the Southern Cross is named Acrux and the two pointer stars are Rigil Kentaurus and Hadar.

If another line is drawn from the midpoint between the two pointer stars at 90 degrees and projected to the first line, the point of intersection is within 1 degree of the South Celestial Pole. This is close

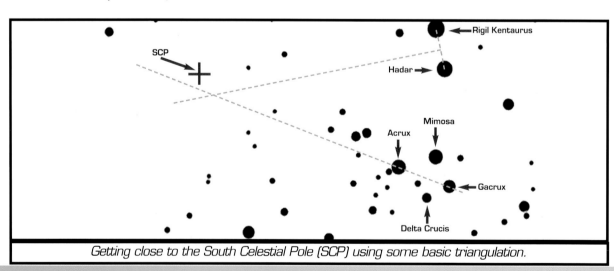

Getting close to the South Celestial Pole (SCP) using some basic triangulation.

enough for viewing, but should be fine-tuned for astrophotography.

Fine-Tuning Polar Alignment (1)

1   The following assumes that you have completed the steps in the previous section, so that your telescope is pointing at the relevant Celestial Pole within about 1–2 degrees.

2   Before you continue, check that your finderscope is properly aligned and looking at precisely the same point in the sky as the telescope.

3   If it isn't, refer to Telescope Adjustments and Maintenance and realign the finderscope.

4   As well as the above, you should check that your declination setting circle is accurate.

5   To do this, make sure that the declination setting circle is set at 90 degrees and locked.

6   Now while looking through the eyepiece, rotate the telescope tube back and forth in right ascension.

7   If the stars appear to rotate around the center of the eyepiece, the declination setting is correct.

8   If they do not, adjust the declination setting slightly and try again.

9   Keep doing this until the stars do revolve around the center of the

eyepiece and then adjust the declination setting circle until it reads 90 degrees.

Fine-Tuning Polar Alignment (2)

1   Ensure by using the methods covered previously in this chapter that your telescope's Polar Axis is pointing towards True North or True South.

2   In addition, ensure that the angle of the telescope's Polar Axis   matches the latitude of your location.

3   Make sure that the declination setting circle is set at 90 degrees. If it isn't, adjust it so that it is and lock both the declination and right ascension axes to ensure they cannot move.

4   Using the illustrations on page 56 as a guide, and utilizing the finder scope (not the eyepiece), adjust the whole mount until the view through the finderscope's crosshairs match es the position shown on the charts for the Celestial Pole you use.

     Please note that you should not adjust DEC or RA in this step, but rather the whole mount.

     When using the finderscope with one eye, it can be helpful to use the other eye to gauge which way to move the mount.

5   Once done, lock the whole mount so that it cannot shift. Your telescope should now be pointing at the North or South Celestial Pole.

This photo of the Moon was taken by the Apollo astronauts from a distance of 10,000 miles on their return to earth. However, you do not need to join NASA to obtain images like this. A combination of telescope and camera will do the job for you from the safety of Earth.

Finding Your Way Around the Night Sky

Now that you know how to polar align your telescope, let's discuss what to do next.

Whether or not your telescope is on an equatorial mount, or whether you are using binoculars, you will still want to know what you are looking at.

The night sky is a vast panorama filled with thousands and thousands of objects that, to the untrained eye, are very pretty but, for the most part, indistinguishable from each other.

So how do you know what to point your new binoculars or telescope at or indeed, if what you think you are looking at is what you think it is?

Acquiring a thorough knowledge of the skies is a very long-term proposition, but if you start with a basic knowledge, you will be able to navigate your way around quite well.

Obviously, the more practice you get the more knowledge you acquire and the quicker and more proficient you will become.

As with many things in life, the best way to start is what looks to be the hard way.

The funny thing is that once you start, it turns out not to be as difficult as you thought. It can even be fun, which is always a help.

There are some aids that can help you along the way. Three of them are as follows:

1   A planisphere, or

2   A bright star atlas (star map), or

3   A computer planisphere and star map program.

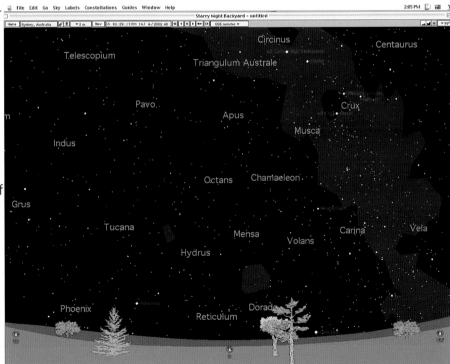

The night sky view from my backyard on June 14, 2002 as shown by a computer planisphere program (Starry Night Backyard). These programs can be adjusted for location, date and time, and are a wonderful way to learn about the night sky.

Note that a planisphere will show what's in the skies at a particular time, but only the objects that are <u>outside</u> our solar system.

To find the movement of objects in our solar system you need an ephemeris.

You can adjust a planisphere for date and time simply by rotating the dial.

A bright star atlas shows the same information, in more detail, over a number of pages.

An ephemeris, which will give you information for objects in our solar system over a year.

But again, for objects in our solar system, an ephemeris is needed.

A computer planisphere program will show both objects in and outside our solar system on the screen, and you can adjust it for date, time and position.

If you want to work with the current positions, you can print them out or, if you have a laptop computer, take it outside with you.

Some computer planisphere programs will connect to certain telescopes and find any object viewable at that time, which you select.

A bright star atlas, which will also assist you to learn the night skies.

The Basic Method (Two Variations)

Variation 1 (Using a planisphere or a bright star atlas to star-hop)

To start, I would suggest that you get yourself a planisphere for your area, similar to the ones illustrated on the next page.

You can do what I am about to suggest using a star map, but I think you will find

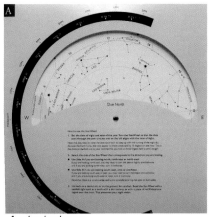
A planisphere

a planisphere much easier to use initially.

Later, you can graduate to star maps, if you desire.

Planispheres are available for both the Northern and Southern Hemispheres.

Remember, a planisphere will only show objects in the skies that are **NOT** in our solar system.

Objects that **ARE** in our solar system are constantly moving across the skies from day to day following the "ecliptic."

The path of the ecliptic is shown on the planisphere as a dotted line.

To find the position in the skies of an object in our solar system, you need to refer to an ephemeris, which is a yearly journal that shows the position of objects in our solar system throughout the current year.

The line on the planisphere that is marked as the equator is the celestial equator, which divides the skies into the Northern and Southern Hemispheres.

The hole in the center of the planisphere represents either the North or South Celestial Pole, depending upon which hemisphere you are in.

One side of the planisphere, when oriented to the south, will show you what is in the sky to the south (ahead), to the east (to your left) and to the west (to your right).

Turn the planisphere over and orient it to the north. Now the northern night skies are ahead of you, the west is to your left and the east to your right.

It could be a good idea to forget your telescope for a little while and use your eyes or binoculars for this exercise.

The planisphere has two positioning devices that need to be set to accurately show what is in the night skies at a particular time.

The first is the hole in the center, which needs to be oriented either north or south, depending upon which way you are looking.

As stated earlier, you need to be using the corresponding side of the planisphere.

The second is a rotating dial, which you can adjust for the date and time.

Adjust the dial to the current date and time. Remember to subtract one hour if your area is on Daylight Saving Time.

Now you need to get yourself out in your backyard on a clear night.

Better still, find a nearby location that isn't exposed to town lights.

Basically, the less lights around you the more you will see, but you do need relatively clear skies, and a red torch.

Locate either the North or South Pole and hold the corresponding side of the planisphere in front of you, so that the hole in the center is pointing towards the pole.

Now look at the planisphere and you will see a map of what is in the skies in the direction you are looking, at that time.

Now look at the sky. What you see in the heavens should match what is showing on the planisphere.

Now start to identify objects, just one to start with. You could pick galaxies or stars, it really doesn't matter.

What's more important is that you are starting to acquire the necessary knowledge t navigate your way around the skies.

In the Northern Hemisphere, you can make it easy on yourself and find Polaris, the North Star.

In the Southern Hemisphere, again make it easy for yourself and locate, say, the Southern Cross (Crux).

Whatever object you pick will be your reference point.

The planisphere gives you the approximate position in the sky, and the surrounding star patterns assist you to zero in on its precise location.

Once you can easily identify a star that is your reference point, use it along with the planisphere to locate three other stellar objects that you know of.

Make sure the objects are showing on the planisphere at that time though,otherwise you will have a most unsatisfactory and frustrating evening.

In the Northern Hemisphere, you might pick Vega, Procyon and the Andromeda Galaxy.

In the Southern Hemisphere, it could be Orion, Canopus and the Corona Australis. Pick large objects at the start.

It really doesn't matter what you pick, as long as you find them and continue your learning experience.

Again, use the pattern of star groups surrounding your next target to help you identify it.

Remember to follow the instructions on the planisphere regarding its use.

Once you are comfortable with the planisphere, you can get a bright star atlas, so that you can start to identify objects using decli-

A laptop computer combined with a planisphere software program is ideal to learn the night skies.

nation (DEC) and right ascension (RA) coordinates.

To do this, you will need to use a telescope that is polar aligned.

Keep adding to your knowledge as time goes by and pretty soon you will have acquired skills that will give you much pleasure in the years ahead.

Variation 2

Instead of buying a planisphere, you can purchase a number of reasonably inexpensive computer programs that will do the same job for you.

The Celestron NexStar 8-inch GPS telescope.

The advantage here is that they not only map the heavens for you, like a star map, but you can also enter in your location, elevation, date and time and right before you on the screen, the present night sky is laid out in all its glory.

If you have a laptop computer, you can take it with you into the field. If not, you can print out a star map at home with the intended targets marked for your night's explorations.

If you use a laptop, set the screen to its dimmest, so its illumination doesn't interfere too much with your night vision.

The Easy Method: Using a GPS-Equipped Telescope

If you have a telescope equipped with GPS (Global Positioning System), then finding your way around the heavens is not a problem, presuming it is set up correctly.

Set up the telescope as usual, key in your target and press *enter*.

After a few whirs and buzzes, your target appears in the viewfinder for you to view. Too easy.

However, if you find your targets this way, you should still take note of their position in the sky and the objects surrounding them.

This way, you will still start to build up a knowledge of the night skies and where everything is, which is always handy.

With a Telescope Equipped with a Computer-Tracking Device without GPS

When you set up a telescope that has a computer tracking device without GPS, it will ask you to pick two stars at least 45 degrees apart and identify them. These two objects will be the computer's reference points.

Assuming you have followed the advice earlier in this chapter on using a planisphere, this should not be a problem.

Once you have identified two objects, the computer can find any other object for you quite easily.

However—there's always a "however"—you need to be reasonably accurate in setting the two reference points.

1   First, make sure that the first reference point is exactly centered using a low magnification eyepiece in the telescope.

2   Now change the low magnification eyepiece to a high magnification one and refine the object's position in the eyepiece to dead center. Only then can the computer confirm the object's identity.

3   Repeat this procedure for the next object which will be the second reference point.

Once this is done correctly, the computer should be able to find and track any object that is in its database and above the horizon.

Using Sky Coordinates and Your Telescope's Setting Circles

Many telescopes now come with computerized, digital setting circles, as well as manual setting circles.

Please be advised that for novice sky watchers this can be pretty difficult, for a number of reasons. Lack of familiarity with the equipment is one. The setting circles themselves is another.

They require extremely accurate adjustment and quickly lose accuracy as the sky moves during the night.

So why bother to mention this in a beginner's book?

Well, if you stick with your new hobby, you will use them eventually , so you may as well get an overview on the subject.

1   To use the setting circles, find an object in the sky that you can readily identify and center and focus the telescope on it.

2   Now consult your reference book (bright star atlas, ephemeris) and find the right ascension (RA) and declination (DEC) setting for this object, at this particular time.

3   Make sure you are still focused squarely on the object in question, and then adjust the setting circles so they correspond with the coordinates from your reference book.

4   Now look up another object in the reference book that is currently in view and get its RA and DEC coordinates.

5   Adjust your telescope to the new coordinates (don't adjust the setting circles, adjust RA and DEC) and the object should be in view, or close by, depending on how accurate you were.

What's Up There?

Now that we have progressed through the equipment side of things, we had better have a brief look at what you can view.

After all, you had better use some of this newfound knowledge, hadn't you?

The sky is a wondrous place with billions of objects that you could not cover in a hundred lifetimes, so it is a good idea to refine what you wish to view right from the start.

In this chapter, I will be showing you various images of objects. To give you an appetite for astronomy, the images I am using were sourced from NASA and the Hubble Space Telescope.

Now you should remember that images from NASA and Hubble have a few distinct advantages compared to what we mere mortals can produce.

First, they have a slightly larger budget for equipment than what the average amateur will spend.

Secondly, the images were taken from space and therefore closer to the subjects.

More importantly, those shots taken from space are not subjected to viewing interference from the earth's atmosphere.

However, don't let their advantages throw you. Given a combination of experience, equipment and viewing conditions, the sights that you can view from the earth will delight you and make your investment of money and time well worth while.

Indeed many of the most spectacular images of objects in the sky were created right here on terra firma by astronomers who, like you, started out knowing nothing about their new hobby.

So, just what is up there to pique your interest?

Well, the list is pretty impressive. Here is a sample.

First you have the **planets** in our own solar system.

The solar system can be divided into two segments, the inner solar system and the outer solar system.

The inner solar system comprises planets that are closest to the Sun. They include Mercury, Venus, Earth and Mars.

The outer solar system comprises the planets that are further away from the Sun. They are Jupiter, Saturn, Uranus, Neptune and Pluto.

The Inner Solar System

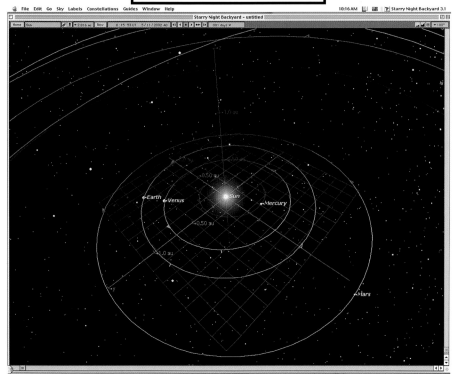

The Outer Solar System

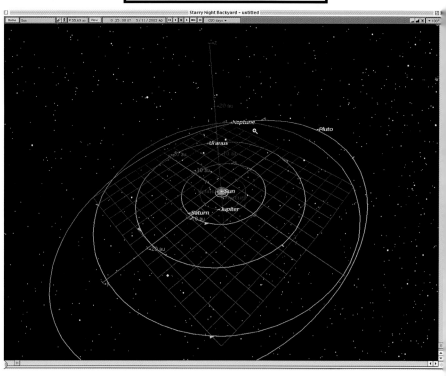

These two images, illustrating the Inner & Outer Solar System, were taken from my computer, using the Starry Night Backyard software, which is one of a number of astronomical software packages available.

As mentioned earlier, this type of software is invaluable in helping the novice astronomer to understand the heavens and also to identify celestial objects. I would strongly recommend you invest in such a software package.

This particular version of the application is the basic one and is relatively inexpensive. More complex versions are pricier, but have distinct advantages as you progress in your knowledge of astronomy.

A montage of the planets, taken from various NASA spacecrafts. Pluto does not appear in this image as no spacecraft has visited it to date. It is interesting to know the diameters of the planets, the Moon and the Sun: Sun 865,000 miles, Mercury 3,031 miles, Venus 7,521 miles, Earth 17,926 miles, Moon 2,160 miles, Mars 4,222 miles, Jupiter 88,730 miles, Saturn 74,940 miles, Uranus 31,763 miles, Neptune 30,775 miles and Pluto 1,430 miles.

So, we have nine planets in our solar system that you can gaze at and photograph, if you wish. But what else is there?

Let's start at the center of our solar system and work outwards.

The Sun is at the center and is the nearest star to Earth.

The next nearest is Alpha Centauri, which is 250,000 times further away.

Because of the Sun's close proximity to Earth, there are **inherent dangers** in observing it, whether with the naked eye or optical equipment.

Obviously, there is the danger to the eyes, but there is also danger to your equipment, because heat can build up in a telescope.

Therefore, you need to ensure that all necessary precautions are taken before attempting to either view (by any means), or photograph, the Sun.

If you are going to use a telescope (either with or without a camera), or binoculars to view the Sun, you must ensure that you have proper solar filters fitted to the telescope's aperture or both apertures of the binoculars.

The filter(s) should cover the entire telescope aperture or, in the case of binoculars, both apertures.

I would also strongly recommend against using a solar filter on the eyepiece instead of the aperture. Always cover the aperture(s).

Also, don't forget to do cover up the finderscope! You should place the plastic lens cap over the finderscope's aperture, so that you don't accidentally use it to view the Sun.

Be especially vigilant if there are children around when you are observing.

Another way to observe the Sun is the projection method.

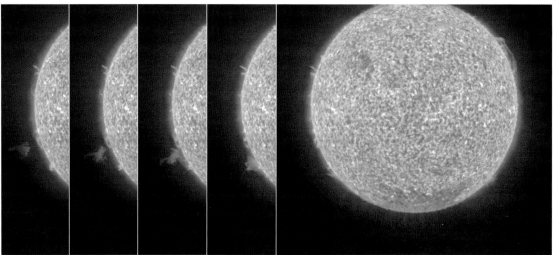

This series of shots of the Sun were taken by NASA in 1996. The eruption on the left hand side ended up measuring 80,000 miles (NASA Image Library).

This has the advantage of costing very little in terms of extra equipment and also allows a number of people to view at the same time.

The projection method involves attaching a metal shaft to the back of a refractor telescope.

Then, using two clips, attach a small white screen and a shade to the shaft.

The shade is there to keep direct sunlight away from the projection screen and sits between the screen and the eyepiece.

Aim the telescope at the Sun, but do not look up the barrel to do this. Instead, position the telescope by looking at its shadow on the ground.

Once aimed correctly, adjust the eyepiece so that the Sun's image is projected clearly onto the screen. You can adjust the shade to better protect the image on the projection screen.

I do not recommend you use your best eyepiece here. A cheaper model will do the job just fine. Also, if your telescope's aperture is greater than around 4 inches, you might consider making a mask to reduce its size.

Please remember the heat buildup problem and keep the viewing times as short as possible.

Give the telescope time to dissipate the internal heat by removing the eyepiece between viewings while, at the same time, fully covering the aperture.

With reflector telescopes, use proper solar filters to view the Sun. However, I have seen the projection method adapted for these types of telescopes.

To view the Sun with catadioptric tele-scopes, you must use correct, full aperture solar filters.

You should, using solar filters, be able to see sunspots and faculae. *Faculae,* which is latin for "torches," are areas of the Sun's surface that are slightly hotter than the surrounding material. They can be difficult to spot.

At an additional cost, hydrogen-alpha filters can reward the viewer by revealing surface details and large prominences.

Picking the right time to view the Sun requires experience and depends on the site. It is worth trying in the early morning.

As with all things, experience is the best teacher, so try different times and locations (by a body of water is good, with the tele-scope pointing over the water).

A solar event that attracts much interest is a solar eclipse.

A solar eclipse occurs when the New Moon passes directly between Earth and the Sun.

Now the Moon is roughly 400 times smaller that the Sun, but it is also 400 times closer to Earth than is the Sun.

These convenient facts mean that, given the right circumstances, the Moon can block the Sun, albeit for a very short time (around 2 seconds up to 7 minutes).

A solar eclipse will not occur with every New Moon, as the orbit of the Moon varies by about 5 degrees, so that the Moon can pass either above, or below, the Sun.

However, when the Moon is in the right place, either a total or partial eclipse can occur.

The shadow cast by the Moon is called the *umbra,* but by the time it reaches the earth, it is fairly narrow, at the maximum, around 170 miles (270 km) wide.

The outer part of the Moon's shadow is called the *penumbra,* and this is where a partial eclipse can be viewed.

A total eclipse is reasonably rare and only visible from certain points on the earth. Between 2002 and 2010, there are only six occurrences.

Generally, in any given year, one or two partial eclipses will occur, but they lack the impact of a total eclipse.

There is one exception, which is called the *annular eclipse.*

This occurs when the conditions are right for a total eclipse but, due to the Moon's or

Earth's elliptical orbits, the Moon doesn't completely cover the Sun, leaving a ring of solar fire (or *annulus*) around the Moon.

Please remember that during an eclipse, you still need to take the necessary precautions to protect your eyes and your equipment's optics.

The planet closest to the Sun is Mercury and we will discuss it next.

Because **Mercury** is the closest planet to the Sun (on average some 36,000,000 miles or 0.39 AU away), it is also a planet that can shine as bright as a star.

However, given that, it can be very elusive to spot.

Apart from Pluto Mercury is the smallest in our Solar System and it has a mass compared to Earth of only 0.06 times.

Its orbit around the Sun lasts only 88 days and it has a very eccentric orbit (only Pluto is more eccentric).

Close up photographs reveal a surface much like the Moon, but without the lava flows that our Moon exhibits.

The best times to observe Mercury are in March or April (evening, Northern Hemisphere) or September or October

Umbra

The umbra defines the area of total eclipse

New Moon

Penumbra

The penumbra defines the area of partial eclipse around the umbra

(morning, Northern Hemisphere).

In the Southern Hemisphere September or October (evening) and March or April (morning)

A closeup of Mercury s surface taken from a distance of 3,340,000 miles in 1974 by Mariner 10 reveals a surface that is similar to that of the Moon, but without the lava flows (NASA Image Library).

The above presumes that Mercury's orbit places it in a position to be visible (called an apparition) at those times.

Venus, sometimes called the Evening Star, as it can be the brightest light in the sky, excepting the Moon, is the next planet out from the Sun

Venus is the planet that comes closest to Earth and also is the brightest planet in the night skies.

In terms of size Venus is approximately 80% of the earth's mass and has a diameter at about 95% compared to the earth.

Approximately 70,000,000 miles (0.72 AU) from the Sun, Venus orbits our Sun once every 225 days. However, as the earth is also moving at the same time it takes

Venus 584 days to reappear in the same position in the sky.

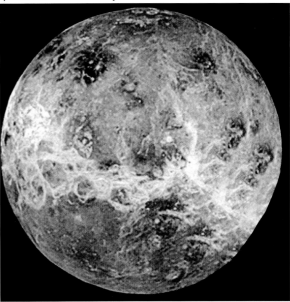

Venus *(NASA Image Library)*

Now you can observe Venus quite easily but it can be uncomfortable, due to the planets brightness.

To counter this try observing at twilight and you will find it far easier on your eyes.

You will find that Venus is rather shy and completely hides its surface under a cover of sulphurous clouds that no normal telescope can penetrate.

These clouds are highly reflective, which is why Venus appears so bright in the sky.

To photograph Venus, try using a transmitting UV light filter such as a Wratten 18A.

Be aware however that if you are using a Schmidt-Cassegrain telescope, its corrector plate will absorb UV light and the result will most likely be unsatisfactory.

Around every 100 years Venus will transit (pass over the face of) the Sun.

The next two occurrences of this are on the 8th of June 2004 and the 6th of June 2012.

Travelling even further away from the Sun the next planet we come to is **Earth**.

As the TV series put it, we are the third rock from the Sun.

Earth is 92,752,000 miles (149,600,000 km) from the Sun and is the most active planet in our Solar System.

Where Earth differs from the other planets lies in the fact that it has an atmosphere (the air surrounding earth) and a hydrosphere (water on or near the earth's surface).

The weather on Earth is caused by the combination of the atmosphere and the heat that radiates from the Sun.

Other factors that influence weather are land and water on the earth's surface and the tilt of the earth's axis (23.5 degrees).

It takes the earth 23 hours 56 minutes to fully rotate and 365.24 days to completely orbit the Sun, hence the need for a Leap Year every four years.

The distance of the earth from the Sun (92,752,000 miles) is used in astronomy as a measuring tool and is referred to as an Astronomical Unit which is expressed as AU. This is always a measure that is calculated from the Sun.

Therefore Earth is 1 AU from the Sun while Mars, being further from the Sun, is 1.52 AU distant.

Telescopes, spotting scopes and binoculars are used on Earth to view terrestrial subjects and have a myriad of applications.

But as well as being the home planet Earth has a near neighbour on which all types of observing apparatus, including our eyes, have been focused on for thousands of years.

That neighbour is, of course, the Moon and is the easiest to observe in our solar system.

The **Moon** is the target of every new astronomer and considering the vast amount of detail that is discernable, it is no wonder.

The Moon is 238,856 miles (384,401 km) from Earth which, in astronomical terms, is spitting distance.

Its orbit, as seen from Earth is 29.5 days and during its travels it presents a number

This photo of the earth was taken by NASA using the GOES 8 weather satellite and was called Big Blue Marble (NASA Image Library)

of what are known as Phases to viewers on Earth.

These phases (there are eight) range from New Moon when the Moon receives no light on the side facing Earth to Full Moon when the entire face is illuminated.

In between, starting from New Moon there is, Waxing Crescent, First Quarter, Waxing Gibbous, Full Moon, Waning Gibbous, Last Quarter, and Waning Crescent. This completes the eight phases of the Moon.

You should also be aware that the face of the Moon that you see is always the same.

This is because the rotation of the Moon is such that it takes exactly the same amount of time as it does to orbit Earth.

The line that divides the visible portion of the Moon from the dark areas is called the *terminator.*

The best viewing of the Moon is not at Full Moon, but rather at First and Last Quarters.

In 1992, the Galileo spacecraft, which was on its way to Jupiter, took this image of the Earth and the Moon (NASA Image Library)

This is because the Sun then strikes the Moon's surface at a shallow angle, which throws the features into sharp relief.

A wide variety of features can be seen at this time, from craters and mountains to lava beds and a feature called *rays,* or ray craters.

Rays are rock that has been smashed by the impact of a meteor and shoot out in all directions from the resulting crater.

Lunar eclipses are another occurrence that is of interest to the avid skywatcher.

Lunar eclipses occur when the Moon passes directly through the shadow that is cast by the earth, which reaches far out into space.

Most times the Moon passes either north or south of the earth's shadow due to the relationship between the tilt of the Moon and the earth's orbit.

However, approximately every six months a Full Moon passes either fully or partially through the earth's shadow.

It is at these times that Lunar (and sometimes Solar) Eclipses occur.

How full the Eclipse will be is dependant on how far the Moon goes into the shadow which is why you can have a Full or Partial Eclipse.

The good thing about Lunar Eclipses is that a far wider section of Earth can see them (compared to Solar Eclipses) and they last much longer.

Now you would expect that if the Moon is fully in the shadow of the earth it would be black, and thus not able to be seen.

Luckily for us, the atmosphere of the earth acts a little like a prism or lens and bends some sunlight into the shadow.

This produces a moon with a red cast, which is clearly visible from Earth.

Binoculars telescopes, spotting scopes and cameras can all be used to view a Lunar Eclipse, as well as our eyes.

Many people photograph Lunar Eclipses and they may take individual shots of each phase of the eclipse or multiple exposures which show all the phases on one frame.

Mars is the next planet outwards and was named after the God of War.

In terms of size, Mars is just over half the size of Earth and it is some 46,000,000 miles away (1.52 AU). The length of a Martian day is similar to Earth (24.6 hours) but it takes 687 Earth days to complete one orbit of the Sun.

Mars has an eccentric orbit and as a result the observer will note that its size and brightness will vary.

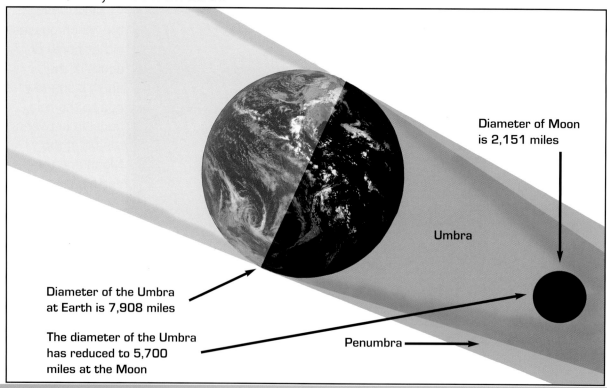

Diameter of Moon is 2,151 miles

Umbra

Diameter of the Umbra at Earth is 7,908 miles

The diameter of the Umbra has reduced to 5,700 miles at the Moon

Penumbra

The best viewing of Mars occurs when the earth sits directly between Mars and the Sun.

This is called an *opposition* and they occur at just over 2 years apart.

Mars apparent size will still vary at each opposition depending upon what part of the orbits of both Mars and Earth the planets are at.

The next opposition is to take place on August 28, 2003 and the planet will be at it largest apparent size, for some considerable time, as its distance from the earth, at that point, will only be some 34,700,000 miles.

Mars as photographed by NASA s Hubble Telescope.
[NASA/STScI Image Library]

To get some good detailed sights of Mars you will need a reasonably large telescope (5 inch or more).

Smaller telescopes will show you something of the planet but you really need bigger to get any detail.

Filters can help with your viewing of Mars and I would suggest you try the following Wratten filters, 44a, 47, 58, 15, 25.

These filters will enhance different features on the planet so just hold them up to the eyepiece and observe.

Observing Mars takes some practice and you will find that your observational skills improve with experience.

Mars does have two Moons called Phobos and Deimos.

However, they are very small and scientists think that they were most likely asteroids that were captured by Mar's gravitational field.

Jupiter, the next planet, is the giant of our Solar System. In terms of size it is 318 times bigger than Earth with a radius over 11 times as large.

In terms of distance, Jupiter is 5.2 AU from the Sun or nearly 390,000,000 miles from Earth.

It takes Jupiter 11.86 Earth years to complete one orbit of the Sun, but the planet rotates at a very fast clip with one full rotation, taking between 9.84 and 9.93 hours.

Astronomers consider Jupiter a border planet as the planets from Jupiter back to the Sun are small and rocky whereas from Jupiter out to Neptune they are large and gaseous (Pluto excepted).

Jupiter has four large moons and (at least) twelve small ones.

The four large moons are Io, Europa, Ganymede and Callisto. Early observations of them led to the first calculations in regard to the speed of light, back in 1675.

Jupiter itself is thought to be mainly hydrogen and helium so what we see is the roof of a very cloudy and deep atmosphere.

One of the most prominent features is the Great Red Spot which is about twice the size of Earth. It was first observed in the 17th Century and has been present ever since, although it's character alters over time.

Two more images from the Hubble Telescope: Jupiter, (above) clearly showing the Great Red Spot and Io (below) . The red markings on Io indicate volcanic activity. (NASA Image Library)

The four bigger moons are also easily visible and provide interesting observing.

To enhance viewing try using Wratten 80A or 82A filters. Also a W12 filter can help bring other details into sharper relief.

Our journey away from the Sun takes us 792,102,000 miles from the earth (19.2 AU from the Sun) to the next planet, which is **Saturn**.

Jupiter is also thought to have a rocky iron core around fifteen times the mass of the earth.

Due to its size even a small telescope or binoculars will present the observer with detail of Jupiter and using larger models the detail becomes even more complete and diverse.

One of the true delights of astronomy is your first viewing of Saturn through a telescope.

Saturn is known as the Ringed Planet.

Now Neptune, Jupiter and Uranus also have rings but Saturn is the only one where you can view the rings with an amateur telescope.

Another large gaseous planet with a small core of dense material Saturn is just over 95.2 times the mass of Earth and it completes one rotation in just over ten hours. To achieve a full orbit of the Sun however takes 29.5 years.

Saturn has some eighteen moons (known to date).

Most of these are beyond the capability of amateur telescopes to view, however this is not the case with Saturn's biggest moon, Titan.

Saturn (NASA Image Library)

Titan has a diameter (3,200 miles) that is bigger than Mercury and can be easily seen from Earth using a telescope of modest aperture. Titan orbits Saturn every sixteen days.

However, it is Saturn's rings that first grab your attention.

The rings are formed by a collection of particles of rock and ice orbiting the planet.

There are three distinct rings named A, B & C ring.

A ring is the one most distant from the planet and is some 9,000 miles across.

Between A & B ring is a dark area called the Cassini Division which was named after the Italian astronomer who discovered it in 1675. It is some 2,600 miles wide.

Originally though to be empty it is now known that this is not the case, there is just much less material there.

The next ring (B ring) is some 16,000 miles and is the brightest of the three.

Finally there is C ring which is closest to the planet and is some 10,500 miles across,

Your viewing of the rings will depend on the planet's position in relation to its orbit. Around every 15 years, the rings can seem to disappear for a time.

To help improve viewing try using a yellow filter in full darkness.

Leaving Saturn, we head further out into the Solar System travelling some 1,780,838,400 miles from the Sun to the planet **Uranus**.

Uranus is another one of the gas giants with an atmosphere that is about 1.3 times the density of water and mainly comprised of hydrogen and helium, with some methane thrown in for good measure.

From Earth, it appears a pale blue-green due to the methane contained in the atmosphere and was initially thought to be a star in the 17th century.

However, this was soon corrected and it is now known that Uranus is some 14.5 times Earth's mass and takes 84 years to complete one orbit of the Sun.Uranus rotates once every 17.2 hours.

Like Saturn and other planets, Uranus does have a ring system which comprises 11 rings and several partial rings, but don't expect to see them from Earth with your telescope.

Uranus does have a number of moons, all named after Shakespearean characters.

Five of these moon, are quite large and are named Titania, Oberon, Ariel, Umbrien and Miranda.

To view them from Earth is quite a challenge, given the great distance involved, but it can be accomplished.

As well as these five there are a further twelve moons that are smaller.

Due to the great distance little was known about the moons of Uranus until NASA sent Voyager 2 by the planet in 1986.

Due to its dense atmosphere details of Uranus are difficult to discern and the images taken by Voyager were featureless.

However, the advent of the Hubble Telescope changed that situation, and new information is being gathered about Uranus each year.

In 2007, Uranus will reach its equinox, at which time it is hoped more detail will be discernable to watchers from Earth.

Further away from the Sun, we come to the second to last planet in our Solar System, **Neptune**.

Neptune is about 2,699,083,200 miles (30.1 AU) away from Earth.

Fortunately for observers on Earth, it is quite bright and very big (17 times bigger than Earth), making observing it reasonably easy to do.

Neptune takes 165 years to orbit the Sun and completes one full rotation every 16 hours, 6 minutes.

Neptune was not discovered until the mid-18th century, although scientists had long suspected its existence, due to irregularities in the orbit of Uranus.

It is thought that Neptune may have a small rocky core surrounded by a deep ocean of water, with an atmosphere of hydrogen, helium and methane.

The methane sits on top of the atmos-

A recent Hubble image of Uranus that shows it surrounded by four major rings and 10 of its 17 known moons. (NASA/STScl Image Library)

phere, and this gives Neptune its blue color.

The planet exhibits ferocious winds that can reach speeds of over 900 miles (1,500 km) per hour.

Many of the features observed on Neptune since it was discovered have changed with the progress of time.

These features include the Great Dark Spot, which was the size of Earth and has since disap-peared. This was first noticed dur-ing the Voyager 2 flyby of the planet in 1989.

Like some of the other planets in our solar sys-tem, Neptune has a ring system (consisting of 5 rings) that sur-rounds it, plus two moons.

Neptune and its largest moon, Triton, are shown in this NASA montage. Triton is the largest of Neptunes 9 known moons, and is about 66% the size of our Moon. (NASA Image Library)

The biggest moon is Triton, which has a diameter of some 1680 miles (2700 km), or roughly two-thirds the size of our Moon.

Although Triton was discovered quite soon after Neptune, the second moon, called Nereid, was not found until 1949.

Triton is the largest of the moons in the solar system to orbit its planet in a back-wards, or retrograde, motion.

The last planet in our solar system, which is quite small and very distant, is **Pluto.**

If you are looking for a planet that we know very little about, Pluto is it.

Pluto was not discovered until 1930, and was found in much the same way as Neptune.

Pluto is approximately 3,570,952,000 miles from Earth, or 39.5 AU from the Sun.

The planet has a diameter of some 1,425 miles, which makes it only 0.002 times Earth's mass.

No wonder it has always been hard to find.

Pluto has a very odd ellip-tic orbit. Indeed, part of it crosses Neptune's orbit and gets closer than Neptune to the Sun.

It takes Pluto some 248 years to complete one orbit of the Sun, and it rotates comparatively slowly, completing one full rotation in 6.39 Earth days.

In 1978, it was discovered that, despite its small size and distance from Earth, Pluto had a moon. It was named Charon.

Separated from Pluto by only just over 12,000 miles Charon is around half the size of Pluto, with a diameter of 760 miles.

Pluto and Charon are sometimes called the "Double Planet."

Pluto does have an atmosphere, which comprises mainly nitrogen and methane.

However, Pluto's gravity is too weak to contain its atmosphere, which, as a consequence, is slowly escaping into space.

As we said before, observing Pluto from Earth is not an easy proposition.

The planet is so small and so far away that it becomes nearly indistinguishable from surrounding stars.

This Hubble shot shows Pluto and Charon, which are sometimes called the Double Planets. (NASA/STScI Image Library)

GPS and computer-driven telescopes can make the job of finding Pluto much easier, as can finder charts that are published in astronomy magazines.

However, even with this assistance, the task is not easy, and it normally requires several nights of observation to spot the small blob of light that is moving among the stationary stars.

Well, we have covered all the planets in our solar system, so that should be it, right?

Not quite, as there are a few more objects of interest that can be viewed.

For instance, there are **artificial satellites** that orbit the earth continuously, including the new International Space Station. and the Hubble Space Telescope.

And let's not forget NASA's Space Shuttle, which is often in orbit as well.

There are hundreds of satellites in orbit and many are visible to the naked eye.

These satellites have a myriad of applications in astronomy, navigation, communications, military operations, earth-resources monitoring and geophysics.

Satellites at a high altitude appear to move more slowly than those in a lower orbit.
There are three kinds of satelllite orbits: equatorial, polar and geostationary, or geosynchronous.

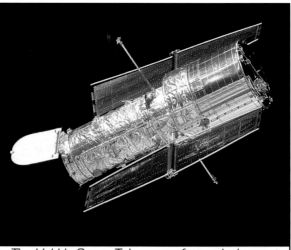

The Hubble Space Telescope after redeployment in 1997, following its second servicing. (NASA Image Library)

The lowest orbits are equatorial, usually at an altitude of around 100 miles (160 km). They travel in a west-to-east direction.

The Space Shuttle Challenger over a very cloudy Earth. This image was taken from the Space Pallet Satellite. (NASA Image Library)

At an altitude of about 1000 miles (1,600 km), you will find the satellites that follow a polar orbit.

These travel either in a north-to-south or south-to-north direction.

Further out in space, you will find the geostationary or geosynchronous satellites. These circle the earth once a day at a altitude

This NASA image of the International Space Station was shot from the Space Shuttle Discovery (NASA Image Library)

of 22,000 miles (36,000 km).

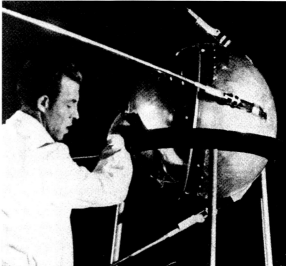

On October 4, 1957, the Soviet Union successfully launched the world's first artificial satellite, called Sputnik 1, which re-entered the earth's atmosphere the same day. (NASA Image Library)

At this height, their speed matches the earth's rotation, and this keeps them over the same point above the earth.

These types of satellites are used for telecommunications, television transmissions and weather forecasting.

For more detailed information on satellites and where to view them, I strongly recommend a visit to the Visual Satellite Observer Website, where there is a wealth of information to be perused: http//www.satellite.eu.org/satintro.html

Comets are another kind of interplanetary objects that can be observed in our solar system. A comet is an object, several miles wide, consisting of frozen rocky or muddy material.

This cold ball consists of water, cyanogen

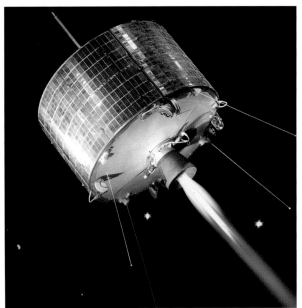

In 1963, the world's first geosynchronous satellite, called Syncom, was launched into orbit. Unfortunately, it failed to send signals after achieving orbit at 22,300 miles (35,900 km). It was soon followed by Syncom 2 and Syncom 3, which were successful and demonstrated the viability of geosynchronous satellites (NASA Image Library)

3,246,320,000 miles from the earth] and 92,752,000,000,000,000 miles away.

Comets formed out of a dense cloud of gas and dust that wasn't incorporated into the planets when the solar system was formed some 4.5 billion years ago.

The gases form a large cloud, or *coma*, which can be thousands of miles in width. It also forms two distinct tails, which can be millions of miles long.

The two tails are an ion tail and a dust tail. Ion tails are formed by ionized gas and appear blue in color. Dust tails are formed by dust and debris and are yellow in color.

When the comet heads towards the Sun, the tails react to the solar wind and fan out behind it.

However, when the comet moves away from the Sun, the tails lead, again reacting to the solar wind.

and what scientists describe as CHON particles, organic material containing carbon, hydrogen, oxygen and nitrogen.

Millions and millions of comets reside in the Oort Cloud, which is about one light year away from the Sun.

Others reside in the Kuiper Belt, which starts just beyond Pluto (about 35 AU or

Comets travel in an elliptical orbit that is very elongated, and once they travel around the Sun, they head out never to return or, at least, not for millions of years.

There are, in fact, two classes of comets.

In this Hubble Space Telescope image, the comet Linear is shown having the comet equivalent of a volcanic explosion. In the second image the explosion occurs and in the third you can see the debris being ejected along the comets tail. (NASA/STScI Image Library)

The first, called short-period comets, like the comet *Halley*, which do return in much shorter time frames (76 years in the case of Halley's comet).

Comets with an orbital period greater than 200 years are called long-period comets. The recent return in 1997 of the Hale-Bopp comet was its first in 4,000 years.

Because it had a close encounter with Jupiter during that trip, its orbit has changed, and it is now expected back in only 2,400 years.

Asteroids and meteors are also present in our solar system and can be observed either with optical equipment or, in the case of meteors, the naked eye.

Asteroids are reasonably large chunks of rock, which are thought to be the remnants of larger bodies that have collided over time, and thus reduced in size and number.

Having said that, there are tens of thousands, perhaps millions, of asteroids that inhabit our solar system. One estimate put the number of asteroids over 0.5 miles (1 km) in diameter at one million.

Their main place of residence is in what is known as the Main Belt, which is between the orbits of Jupiter and Mars.

The largest asteroid in the solar system, nearly 600 miles (1,000 km) wide, is called Ceres.

Many asteroids are too small to be seen from Earth with a telescope, but that still leaves several hundred that can be spotted and tracked.

Most astronomical software applications include asteroid positions and, furthermore, astronomy almanacs publish the positions of the brighter ones.

The collisions that sometimes occur between asteroids casts off fragments, and some of these fragments will cross Earth's orbit as meteoroids.

Indeed, asteroid fragments are the main source of meteoroids, the other being the debris that comes off of comets as they travel through the solar system.

If a part of the meteoroid survives the atmosphere and lands on Earth, it is called a meteorite.

Ilt is estimated that 4 or five enter our atmosphere every hour, but they generally burn up before they reach the ground.

Larger meteoroids that explode on contact with the earth's atmosphere, generating brilliant balzes of light, are called fireballs. They usually last a few seconds, leaving a smoke trail behind.

Meteoroids can appear to come from one sector of the sky that is called the radiant.

Occasionally, the number of meteoroids entering our atmosphere increases substantially, especially during meteor storms, which are rarer than meteor showers.

You do not need telescopes or binoculars to view meteors, but how do you know when they will occur?

Every night there will be some meteor activity in the sky, but how well you can see it depends on your location.

The best conditions are when the night is darkest and the sky is clearest.

This means that you must get as far away from city lights as possible, especially because many meteor trails are quite faint.

Meteor showers will most likely be predicted in the news and meteor storms will most certainly be broadcast.

There are ten meteor showers that occur each year and these are detailed on the table on the next page.

Normally, the hours between midnight and dawn are the best time to watch a meteor shower.

Well, that should give you enough to look at in the solar system for a start.

In the following pages, we will go further afield, outside our solar system, and then have a brief look at various equipment that can be used to view them.

Serendipitously Discovered Asteroids
Hubble Space Telescope • WFPC2

Astronomers have started to comb the vast image library of the Hubble Space Telescope. When Hubble trains its telescope and cameras on an object in space, the exposure made can last a very long time. Any asteroid that happens to come into the field of view of Hubble during this period will be recorded on the image as a bright, curved blue line. This line traces the path the asteroid took as Hubble was exposing the image. The images above are examples of this, and they and others have been responsible for the identification of many previously unknown asteroids. (NASA/STScI Image Library)

When observation extends outside our solar system, it is known as deep sky observation—and what a fascinating pastime it is.

You can use your eyes or binoculars for some objects, while others will require a telescope with an aperture of 4 inches or more.

Before going further, we had best discuss apparent magnitude and absolute magnitude

Apparent Magnitude is a system based on logarithmic unit that defines the optical brightness of a celestial object.

An object with a lower magnitude is brighter than one with a higher one. For instance, magnitude 1 is 100 times brighter than magnitude 6. The difference between, say, magnitude 1 and 2 is that

A Guide to Predicted Annual Meteor Showers			
Shower Name	Activity Period	Peak Times	Est. rate per hour
Quadrantids	Jan 1 – 5	Jan 3	40 – 100
Lyrids	April 16 – 25	Apr 22	15 – 20
Eta Aquarids	Apr 19 – May 28	May 4	20 – 50
Delta Aquarids	July 8 – Sept 20	July 29	20
Perseids	July 17 – Aug 24	Aug 12	50 – 100
Orionids	Sept 10 – Oct 26	Oct 22	25
Taurids	Sept 15 – Nov 15	Nov 3	12 – 15
Leonids	Nov 14 – 21	Nov 17	10 – 15
Geminids	Dec 7 – 17	Dec 14	50 – 80
Ursids	Dec 17 – 26	Dec 22	10 – 20

The faintest object that you can see from the earth with the naked eye has an apparent magnitude of 6. Therefore, any object that has an apparent magnitude of less that 6 (5, 4, 3, 2, 1, 0, -1 and so on) will also be visible.

The table on the next page gives the apparent magnitude of the Sun and the planets.

Please note that the variations in magnitude are due to the relationship of the planet concerned with Earth's position.

In addition to the Sun, there are four other stars that are brighter than 0.

Apparent magnitude therefore gives you a good indication of how easy or difficult an object in the heavens will be to view.

Look up at the sky on a clear, moonless night. What do you see?

Stars—trillions of them. Some shine very brightly, while others are dim specks of light.

From Earth, you can observe stars with your eyes, binoculars or telescopes.

Stars come in different sizes and colors. For instance, the 200 stars mentioned in the caption for the image on this page are 15 to 60 time bigger than our Sun.

Stars appear to observers on Earth in different colors, from white to red, depending on their temperature.

Stars that emit white light are very hot and very bright, and are known as White

magnitude 1 is 2.512 times brighter than magnitude 2.

Put simply, the difference between apparent magnitude and absolute magnitude is this:

Apparent magnitude is how the object appears when viewed from Earth.

However, the objects we are discussing are all at different distances from Earth, and it is possible that an object that looks brighter than another object may, in fact, have less optical brightness. But because the second object is much further away, it looks dimmer.

This is where absolute magnitude comes in. It eliminates the effect of distance on an object's magnitude by calculating its brightness as if it lay at a fixed distance of 32.6 light-years from the Sun.

Since it is from Earth that we view objects, apparent magnitude is of more interest to us.

The Sun and Planets	
Object	Apparent Magnitude
Sun	-26.7
Mercury	-2 to +3
Venus	-4 to -4.6
Moon	-12.7 (Full Moon)
Mars	-2.6 to +1.8
Jupiter	-1.2 to -2.5
Saturn	0.6 to 1.5
Uranus	5.5 to 5.9
Neptune	7.9
Pluto	13.7
Alpha Centauri	-0.01
Arcturus	-0.04
Canopus	-0.07
Sirius	-1.5

Not including the Sun, stars are located about 4 light years away up to truly incomprehensible distances.

Stars do not appear to move when observed from Earth.

But the stars do move. It's just that it is very difficult to spot the movement.

Even our the Sun is moving towards the star Vega at a rate of 43,000 mph (72,000 kph).

The shift in a star's position is called its proper motion, and it is normally quoted in units of seconds of arc per year—which is a very small movement, when viewed from Earth.

Dwarfs. The red stars are much cooler, but also much bigger, and are heading towards the end of their life-cycle. They are known as Red Giants.

Stars come in a variety of packages.

For instance, there are Double Stars and Variable Stars.

Each of these types also has its own sub-categories, so we had best take a closer look at them.

There are two types of **Double Stars**. First, there are true binary or multiple star systems.

A Nursery of New Stars.
This image, taken by the Hubble Space Telescope, is a vast nebula called NGC 604 in a neighboring spiral galaxy M33, which is a mere 2.7 million light years away in the constellation Triangulum. Such nebulae are common in the galaxies, but this one is particularly large, being almost 1,500 light years across. New stars are being formed in the galaxies spiral, and this is illuminated by over 200 huge hot stars in the heart of the nebula. (NASA Image Library)

These systems can contain two or more stars. They occur when the stars are in an orbit around a common center of gravity.

Another type of double stars are called Optical Doubles.

These stars are not true doubles, but appear thus due to the apparent closeness of their alignment when viewed from Earth. In reality, they can be many light years apart, and they certainly do not share a common center of gravity around which they orbit.

Observing Double Stars can be interesting and extremely challenging, precisely because of the optical illusion and the great distances involved.

The second class of stars we mentioned are **Variable Stars**.

These are stars whose brightness varies over time. The time period involved can be short or long, regular or irregular.
There are three general classes of Variable Stars

The first is known as a **Pulsating Variable.**

These are stars that regularly fade and brighten either over short or long periods of time, ranging from one day to several days.

Red Giants are Variable Stars that can pulsate from around 80 days to 5 years.

The second type of Variable Star are called Eclipsing Variables.

These are binary, or double, stars that,

due to their orbit, regularly eclipse each other.

When viewed from Earth this manifests as a regular decrease in light output which returns to normal until the eclipse occurs again.

The third category are known as **Cataclysmic or Eruptive Variables.**

Many are binary systems that can have a

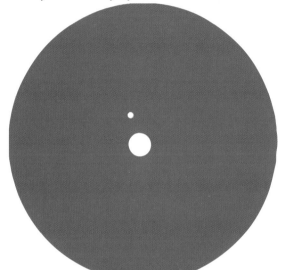

A two (above) and three (below) Double Star system.

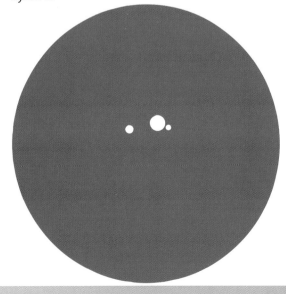

Variable Stars in a Distant Spiral Galaxy.
This Hubble Space Telescope image shows the distant spiral galaxy NGC 4603.
This galaxy has a special class of pulsating variable stars called Cepheid variables, and this is the most
distant galaxy where they have been found. (NASA Image Library)

periodic outburst of brightness that may last many thousands of years.

Others are so violent that they destroy the star and are known as *Supernovae*.

Stars are found in **Galaxies** and Galaxies are objects that can be observed in their own right. Galaxies are a collection of billions of stars, gas and dust that are held together by gravity.

Galaxies range in size and shape. In terms of size they can be a modest 1,000 light-years to 100,000 light-years across.

As far as shape is concerned they are categorized into three main classes, which are spirals, barred spirals and ellipticals.

Observing Galaxies can be difficult as it depends on how the Galaxy presents to

Earth. It can be face on where we can view the entire width, or edge on where we are only seeing the leading edge of the Galaxy.

Some Galaxies can be seen clearly from Earth with the naked eye, for instance the Large and Small Magellanic clouds are clearly visible in the night sky in the Southern Hemisphere.

Other Galaxies require telescopes to view them (from 4 inches up is good), but the effort will be well rewarded.

As with many things in life, the better the equipment used, the better the image that is presented to the viewer. Unfortunately, this is always at a increased cost.

Stars are made up from gas and dust and these materials are found in **Nebulae** throughout the heavens.

As stars form in Nebulae they can become bound to each others gravity and form clusters that, as they move over the eons, will eventually be dispersed by the influence of gravity from other bodies.

There are two categories of Nebula being dark and bright.

Dark Nebula are clouds of dust and gas that are dense enough to block the light from stars that are behind them.

Bright Nebula fall into two sub-cate-

gories, which are Emission and Reflection Nebula.

A Reflection Nebula is one where the light that is seen is simply that which has been generated by a nearby star and is being reflected by the Nebula's dust.

An Emission Nebula generates its own light. What happens is that a nearby star ionizes the atoms within the dust and they start emitting their own light.

There are also Nebulae that are a

In 1995, and again in 1999, the Hubble Space telescope composed this magnificent image of the majestic spiral galaxy NGC 4414.
The image shows that the central regions of this galaxy contain mainly older red and yellow stars. This is typical of spiral galaxies. However, the outer spirals are considerably bluer due to the fact that they contain much younger, recently formed stars, a process that continues today. The galaxies arms are full of very rich interstellar dust, which shows as dark streaks and patches. (NASA Image Library)

combination of both and these are known as Emission/Reflection Nebulae. The colors of a Nebula will vary depending on how you view them.

For instance, if you are looking at a Nebula through a telescope you will not see color as our eyes are not sensitive enough to detect it.

By this stage you have a fair idea of what is up there for you to view but what piece of equipment (telescope or binoculars) is best for what situation?

Let us look at the planets and moons within our Solar System first and see what the three types of telescopes can do as well

Close Encounters of the Galactic Kind.
Two spiral galaxies, NGC 2207 (left) and the smaller IT 2163 (right) had a close encounter around 40 million years ago, according to current scientific calculations. The larger galaxy, with its stronger gravity is distorting the smaller one, which is not strong enough to escape its bigger brother. Some time in the future, they will merge into one bigger galaxy.
(NASA Image Library)

However, film and CCD image sensors are sensitive enough to detect color, and you will find that photographs of Reflection Nebulas are a blue shade while Emission Nebulas are red.

Dark Nebulas are harder to spot and are most visible if the are silhouetted against a bright nebula.

The first question should always be what do I want the equipment to do?

Do you want to look at only the planets and moons within our Solar System and leave deep sky observing alone, at least for the moment?

The second question, a far more practical one, is what will my budget allow?

Two Galaxies NGC 3314 (front) and NGC 3314b (back) that the Hubble Space Telescope studied in 1999 and 2000. The front galaxy is face on, while the background galaxy is edge on and tilted. However, because the rear galaxy (a spiral) is much larger, it forms a perfect backdrop to see detail in NGC 3314. The latter is also a spiral and is located 140 million light years from Earth near the constellation Hydra in the Southern Hemisphere. (NASA Image Library)

You will remember that there are three basic telescope types being the refractor, the reflector and the Schmidt-Cassegrain and Maksutov-Cassegrain variations.

All three of these telescopes will handle the Solar System just fine and you should review the earlier chapter titled "Telescopes How They Work And What To Look For" to refresh your memory as to their advantages and disadvantages when compared to each other.

Reflector telescope on an Dobsonian mount

Schmidt-Cassegrain telescope on an altazimuth mount

Refractor telescope on a Altazimuth mount

Also, don't forget that binoculars also work well here.

Admittedly, they will not get you as close as a telescope, but they have many other advantages, including a wider field of view, ease of use and less weight.

Now if your ambitions go further than our Solar System and include deep sky observation, then the situation is a little different.

Let's face it, there is enough distance to cover inside our Solar System. Add the rest of the viewable universe and you are talking about staggering distances, so you will need as much resolving power as you can afford.

Please don't forget that with telescopes the initial resolving power is determined by the aperture, and not the eyepiece.

The eyepiece simply magnifies the image presented at the telescope's focal plane, by the aperture.

Also remember that an increased power

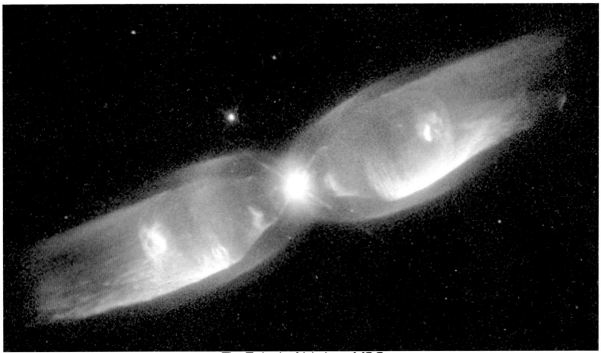

The Twin Jet Nebula or M2-9

This nebula is a striking example of a butterfly or bipolar planetary nebula. As the nebula is sliced across a star, the result is similar to two exhausts of gas from a jet engine, which is where it picks up its popular name. The nebula is located in the Ophiucus constellation some 2,100 light years away. The image was taken by the Hubble Space Telescope in 1997. (NASA Image Library)

eyepiece is no substitute for large apertures.

In the case of deep sky observation, refractors are a tad limited, due to the high cost of the ones with large apertures.

Both reflectors (including Dobsonians) and Schmidt-Cassegrain telescopes are better here due to their affordable larger resolving power.

Both are equally good for observing the planets in the solar system, although I would have to say that a good refractor telescope can give a slightly superior image here.

You should also remember if you plan to take photographs with your telescope that

many objects in the heavens move across the sky from east to west.

Therefore, you are going to have to track the object as it crosses the sky to get a correctly exposed image.

This is because, in most cases, anything you may wish to photograph in the night sky puts out a small amount of light and this necessitates long exposures.

This is covered in more detail in the next chapter, entitled "Astrophotography."

Reflector telescope on an Dobsonian mount

Schmidt-Cassegrain telescope on an altaz-imuth mount

Refractor telescope on a Altazimuth mount

Astrophotography

Astrophotography is a challenging pursuit that involves capturing still images of celestial objects.

Remember that if you are going to photograph the Sun, you need appropriate filters, either for the camera or the telescope, to protect your eyes and equipment.

Anyone who has even owned a camera has most likely pointed it at the sky at one time or another and shot off a few frames.

Likewise, anyone who has a telescope will, at one time or another, get the urge to attach a camera to it and shoot away, so they can share with family and friends the wonders they discover in the cosmos.

Now, astrophotography is quite an exacting discipline and entire books are written on the subject. They can be quite daunting to the novice astrophotographer.

That being the case, how am I going to cover this complex subject in a single chapter of this book?

Well, I'm going to get you started with some simple techniques and pointers. Hopefully, the results you get will encourage you to delve further into this fascinating pastime, and to learn more as time goes by.

We will look at three techniques for taking pictures of the night sky.

As long as you have an SLR camera body, a tripod and a cable release, along with your telescope, you will be able to try all three of the techniques.

A compact camera will allow you to use techniques one and two, but will preclude technique number 3.

SLR cameras are the best choice for astrophotography, but compact cameras can be used in certain situations.

An automatic tracking device for your telescope would also be a definite plus.

Before we look at these three techniques, let's examine some common concerns.

In general, when it comes to photography, there are a number of issues that the photographer must sort out to get that perfect image. These are:

1   Where do I focus to get the best image? Is my "Depth of Field" (see Glossary) important?

2   What ISO rated film or setting (digital cameras) should I use?

3   What aperture and shutter speed combination should I set?

4   How do I ensure that there is no camera shake, which will result in blurred images?

5   Should I shoot color or black and white?

6   How do I get a sharp image (if that's what you want) of a planet or star, which is in constant motion and has a low reflectance of light, requiring long exposure times?

Before we go into the various methods of doing astrophotography, let us attack these problems. It is a certainty that some, or all, of them will affect you.

Issue #1: Focus

If you're using your camera lens to record the image, you will need to set the focus to infinity (∞).

Any object in the heavens is so far away that even if you used the longest focal length camera lens, you will still need to set it at infinity (∞).

When you use a camera in conjunction with a telescope, the telescope becomes the camera lens.

In this case, the problem is not how to focus the camera, but how to focus the telescope.

If your camera is attached to the telescope, where the eyepiece normally resides, how do you achieve focus?

On the surface, it should be simple, but in reality it isn't.

Logic would suggest that when you are using an SLR camera on your telescope (remember, compact cameras won't do the job here), you would simply look through the camera's viewfinder and focus the telescope.

Unfortunately, this can sometimes be a problem. With really big and bright objects, such as the Moon or one of the planets, there should be no problem.

But if you want to shoot a faint object that requires a powerful eyepiece to be seen, there definitely will be a problem.

That problem, and one of the solutions, is discussed later in this chapter.

Issues #2 and #3: ISO Ratings, Aperture and Shutter Speed Settings

ISO ratings refer to the speed of the film you are using or, in the case of digital cameras, the ISO number you set for the CCD or CMOS, two different kinds of imaging sensors, presuming it is adjustable.

In photography, the higher the ISO setting, the less volume of light is required to correctly form the image, either on film or the image sensor.

Every time an ISO number doubles, for instance from ISO 100 to ISO 200, the volume of light required, to correctly expose an image, is halved.

By volume of light, I mean a combination of the amount of light to strike the film or image sensor, controlled by either the camera's or telescope's aperture, and the time it is allowed to do so, which is controlled by the camera's shutter.

You need to remember that in astrophotography, the amount of light that is being reflected from any object in the heavens (with the exception of the Sun and the Moon) is minimal and, therefore, long exposure times will definitely be necessary.

Combine these issues with the fact that your subject is constantly moving, in which case, theoretically, the shorter the exposure time the better.

In reality, this is not necessarily the case.

You see, there are many objects in space that are too dim for the eye to see, at least initially, but will record on camera because of the long exposure times necessary.

You can see this yourself by simply looking at the heavens at night with the naked eye.

If you stare at one spot, you will see some stars. But continue staring at the same area and other stars will start to show up that you could not see initially. Try it.

In addition, there are certain colors of the Northern and Southern Auroras that the naked eye cannot see, but will also be picked up on camera.

So what ISO number should you use?

With Camera Lenses

With lunar photography, you have quite an amount of light to use, so the ISO number you use can be reasonably low.

Try ISO 100 at an aperture of 5.6 and a shutter speed setting of 1/500th of a second for a full moon. For a quarter moon, try 1/30th, a half moon 1/60th and a three-quarter moon 1/250th with this ISO speed.

These settings should get you into the ballpark and you can make adjustments from there.

With other celestial bodies, such as stars and constellations, I would suggest you start out with ISO 400 or 800, if a crisp photo is your objective.

Over ISO 800, digital sensors may run into computer noise. But with film cameras, higher ISO speed film is available, and it is advisable to choose it over slower film.

When trying to attain the optimal exposure times using the largest aperture available on a camera lens, trial and error is absolutely necessary.

Additionally, it also depends on the ambient brightness of the night sky at the place where you choose to take the photographs. You could bracket your exposures.

Bracketing means you take a number of shots at different exposure settings.

Since it would be a good idea to allow as much light into the camera lens as possible, you need to use the largest aperture possible for your lens (equivalent to the smallest f-stop number).

Since you have already set your ISO number, the only other variable is the shutter speed setting for your bracketed shots.

Actually, this is not entirely accurate, for with some digital cameras, you can vary your ISO number (camera permitting) from shot to shot, but here we'll presume you can't.

You also have to consider whether you want your photograph to include star trails, a product of the movement of the stars in the sky.

To avoid star trailing, exposure times of 80 seconds or less are recommended, unless your camera is mounted on a telescope with automatic tracking, which is discussed a little later.

If you don't mind star trailing—and, let's face it, these types of images can be very impressive—then exposure times can blow out to many minutes or hours.

It depends on what you want and where you are, so you'll need to experiment a lot to figure out how best to achieve your aims.

With Your Telescope as the Cameras Lens

Telescopes do not have adjustable apertures, as do cameras, so whatever light gathering ability your telescope has is what you are stuck with.

For instance, most Schmidt-Cassegrain telescopes have an aperture of around f10, which allows in considerably less light that does, say a camera lens set at f2.

Therefore, you are going to need longer exposure times than you would if you were taking photographs with your camera alone.

Now, if you want pin-sharp images, and your telescope has a mount with automatic tracking, it's pretty simple, provided the telescope is properly set up.

You can use an ISO setting of between 50 and 400 and track your subject across the night sky as you photograph. Other conditions being favorable, you'll get pin-sharp images.

If you don't, then the problem is either in the telescope's alignment, levelling or collimation. It can also be the fault of your automatic tracking device, which may require training (see later in this chapter).

Taking lunar shots is easier, even if your telescope is not fitted with automatic tracking.

Try using a film or image sensor setting of ISO 400 and exposures of between 1 second (for a full moon) to 4 seconds (for a quarter moon). Adjust these settings as required when you view the images.

Issue #4: Elimination of Camera Shake

The movement of the camera during

exposures, which have a shutter speed that is slower that the focal length of the camera lens, causes camera shake.

Let's say you have a camera with a 100mm lens attached. Any shutter speed slower than 1/100th of a second increases the risk of camera shake which, of course, can result in blurred images.

As all astrophotography should involve the use of a camera that is mounted on a tripod or, alternatively, a camera mounted on a telescope that is itself mounted on a tripod, you may wonder why I am going on about camera shake.

Doesn't the use of a tripod eliminate camera shake?

No! Sorry, the use of tripods only helps to minimize the risk of camera shake. There are other factors that can cause camera shake.

By the way, the word *tripod* indicates three legs.

If you use the center post of a tripod to hold a camera, or worse, a tele-scope with a cam-era attached, you are effectively using a monopod, which means one leg.

You may be forgiven for think-ing your equip-ment is taking rock-and-roll lessons.

Don't use the center post of a tripod unless you are using fast shutter speeds. You won't like the results, I promise you.

The next thing that can cause camera shake is your finger, as it trips the shutter.

Do not use your finger. Use a cable release that is equipped with a lock that allows the shutter to be kept open for a long as is necessary.

A cable release is another essential piece of equipment for astrophotography.

Another factor to consider when it comes to the camera shake produced by SLR cameras, is the camera's mirror.

The mirror in an SLR camera is there to divert the image into the top of the camera, which is called a pentaprism, and on to the viewfinder, so you can see what you are looking at.

When you trip the shutter, the mirror flies up and allows the light to travel directly to the film or image sensor.

This action is what causes the viewfinder to black out while the shutter is open. Once the shutter closes, the mirror drops down again so you can see through the viewfinder once more.

This action of the mirror flying up can cause camera shake.

A tripod is essential for astrophotography. However avoid raising the centre post if it has one.

To safeguard against this many SLR film cameras have what's called a Mirror Lock-up switch.

Before you trip the shutter, you activate this switch, and the mirror locks up out of the way.

You then trip the shutter for the required time period and when it is closed, you release the Mirror Lock-up switch.

This will ensure that the mirror does not cause any camera shake.

If your camera does not have Mirror Lock-up switch, the only thing you can do is ensure it is mounted as securely as possible. Another factor is the wind.

If you have a camera on a tripod or attached to a telescope, you have a reasonably-sized piece of equipment hanging out there in the breeze.

What happens if it is windy? All that breeze is bouncing off your camera or telescope, and it's rock-and-roll time again.

If this happens, you need to protect the equipment with some barrier that will keep the wind away from the equipment.

If you can't protect the equipment from the wind, I'd return another night when it's calmer.

Issue #5: Black and White or Color?

If all the problems with astrophotography were this simple, I would be lucky to fill one page of this book with this chapter.

Shoot your images in whichever medium you prefer.

Personally, I prefer color. The reason is that color images bring out hues in the night sky and the objects therein that you don't see with the naked eye.

But if you prefer black and white, use it.

Issue #6: Tracking Moving Objects

So you spent all that money on an automatic tracker and it keeps wandering off course ever so slightly, giving you blurred images.

Before you go back to the shop and start yelling at the sales assistant, you might want to look at a few things that may correct the situation.

First, is your telescope properly set up and polar aligned at the start?

If you're only observing stars, an accuracy of plus or minus 4–5 degrees is normally good enough.

Unfortunately, the same latitude does not apply to astrophotography. Here you will need an accuracy of plus or minus 0.2 degrees.

The exception to this is when you're using a fully computerized telescope with GPS (Global Positioning System) tracking capability. But let's go back to the real world for a moment where you don't have a GPS system.

Then you need to be especially vigilant when setting up for your automatic tracker to function. Sometimes, it's the simple things that mess you up.

For instance, you are out in the field and have been especially careful to set up and align your telescope, and still the automatic tracker does not perform correctly and your images are blurred.

Get your red flashlight out and have a good look at the tripod legs.

Do any of them appear to have sunk a little into the ground since you first set the tripod up?

Remember, the tripod carries its own weight, the weight of the telescope and tracker unit, plus the camera.

That's quite a load, and it's surprising that people often overlook this factor when trying to get the automatic tracker to work.

Lastly, who has ever bought a piece of machinery that behaves flawlessly? Not me, and I presume you haven't, either.

With automatic trackers, the main problem is that they don't keep absolutely perfect time with the earth's rotation. This is usually due to minor manufacturing flaws.

To counter this, you need to train the machine to account for its built-in flaws, and to do this you need to look at the documentation that came with the machine.

Unfortunately, the chances are that you will have to do this each time you use the tracker, as they are not built with a memory feature.

Basic Methods of Astrophotography

As I stated at the start of this chapter, I only intend to discuss the three most common methods of astrophotography in this book.

I strongly suggest that you start your with the first method and then progress to the second and third methods.

This will not only give you valuable experience, but will save you money and time by teaching you what you need to know to successfully use the third method.

Note that there are variations to these methods, which you will have to investigate in depth in order to do them justice.

There are some equipment needs associated with these methods, which we will look at now.

Camera. Obviously, you need a camera. But what type?

Well, any camera can be pointed towards the heavens to take photographs so a compact or rangefinder camera or 35mm single lens reflex (SLR) will do fine.

However, to use method 3, the camera must be attached to the telescope, which requires you to use an SLR camera.

This is because with a compact or rangefinder camera, you cannot change lenses, and therefore cannot attach the camera to the telescope.

There are exceptions to this, but the cameras involved are normally pretty expensive, so I am ignoring them here.

In addition, the camera should have a bulb (B) shutter setting. This setting allows the shutter to stay open as long as the shutter release button is pressed down.

The camera can be a film or digital type.

One word of warning: Most modern cameras run on batteries, and long exposure times can drain batteries pretty quickly, so make sure you have plenty of power in your camera before shooting that long exposure.

Many astrophotographers prefer to use traditional mechanical cameras, which eliminates the problem of batteries.

Make sure you have plenty of power for your camera when shooting the heavens.

Next, you need a **cable release** with a lock.

This attaches to the camera and allows you to fire the camera's shutter without direct contact with the camera body. It also keeps the shutter open for extended periods of time.

You may have problems with this if you are using a compact camera, because with a few expensive exceptions, compact cameras are not equipped to use a cable release.

If the camera is not mounted on the telescope, you need a good sturdy **tripod** to hold the camera steady while the exposure is being made.

Please remember the earlier caution against using the center post of a tripod to hold a camera or telescope.

With those few requirements, you have the basic equipment needed to get you started on your adventure.

Now for the three methods I promised to discuss. They are:

1   Shooting from a camera that is mounted on a tripod.

2   Shooting with a camera mounted on a telescope, but using its own lenses. This is known as the piggyback method.

4   Shooting with a camera mounted on a telescope, but using the telescope's optics as the lens.

The Camera Mounted on a Tripod

Any type of camera can be used to photograph the heavens.

You also need a good tripod and a cable release. Your camera should also have a bulb (B) shutter setting.

If you use a digital camera, then you can simply preview your subject on the camera's

LCD (Liquid Crystal Display) screen and reshoot, if necessary.

Some beginners shoot almost exclusively with black-and-white film and process it themselves, which can save time and money in the learning stage.

Regardless of what camera you use, keep a record of each shot, so that you will remember what you did and the conditions at the time.

The basic way to shoot is to set up the camera, trip the shutter and leave it open for a specific period of time.

This will give you star trails, which can make for a spectacular photograph. The trails can be quite short or long depending on how long the shutter is left open.

If the camera is aimed at the North or South Celestial Poles, you will get star trails that form circles, which are also quite impressive.

Most other objects in the sky can be captured on film, but because of their movement, if you want sharp, crisp images, you need quick shutter speeds and wide open apertures.

Exposure settings will vary widely depending on a number of factors including equipment, atmospheric conditions and subject matter. You need to experiment and try various combinations.

Some guidelines:

1   The darker the sky the better.

2   When shooting with a camera, the longer the focal length of the lens, the sooner the stars begin to trail on the image.

3   Stars near either celestial pole take longer to trail that those near the celestial equator.

The Piggyback Method

Shooting with a camera mounted on a telescope, but using its own lens, is the next step up.

The advantage it offers is that, if your telescope has an equatorial mount or, better yet, an automatic tracking device, it can be used to help you track your subject during long exposures.

You can purchase reasonably priced camera brackets that will attach the camera to the telescope tube.

The tips in the previous sections about using your own camera still apply here.

The difference is that you now have a medium to track your subject.

Theoretically, you should now be able to

An SLR camera mounted piggyback on a Schmidt-Cassegrain telescope on an equatorial mount with automatic tracking.

track that faraway object perfectly during a long exposure and get a pin-sharp image.

This is provided that your mount is aligned perfectly and your tracker is behaving flawlessly. There is always a catch, isn't there?

Any flaw in your telescope's alignment or in the tracking mechanism can produce faulty images, so these areas need to be looked at very closely.

Having said that, short duration exposures and the use of short focal length lenses will help cover up minor imperfections in alignment and tracking.

Long exposure times and long focal length lenses will not!

Using the Telescope's Optics as the Camera Lens

This is most probably how most newcomers to astrophotography envisage taking pictures of the stars and planets.

An SLR camera with a camera adaptor to connect to the telescope is required.

Assemble all this equipment and go to the location designated for the shoot, with spare film and batteries.

Next, set up your telescope correctly, so that it is perfectly level and aligned.

Now, find a star that you wish to photograph. It's a long way away but your telescope with its high power eyepiece (160x) can handle that easily.

So out with the eyepiece and in with the camera and its adaptors.

You look through the cameras viewfinder and what do you see? Nothing. Zip. The blasted star has disappeared.

It only took 3 seconds to replace the eyepiece with the camera, so the star hasn't had time to move out of the field of view. It should still be in the telescope's field of view.

So, you take the camera out and reinsert the eyepiece to find that blasted star and there it is. But every time the camera is attached to the telescope, it disappears.

What is going on here?

Well, what's going on is that you have temporarily forgotten how telescopes work.

Earlier, we said that the telescope collects the light being reflected from a star and presents it at the image plane for the eyepiece to magnify and view.

The telescope has its own magnifying power, which is separate from the eyepiece being used, and a factor of its focal length.

Let's say that your telescope has a focal length of 2000mm. This means that it has a magnifying power of approximately 40x (see page 28).

At the image plane is your image of the star, magnified 40x. Then your eyepiece magnifies that image by a factor of 160x.

Now you take the eyepiece out and replace it with the camera.

What you have done is remove the 160x magnification factor and replaced it with a camera body with a magnification factor of 0x. Goodbye 160x magnification. Goodbye star.

Remember, the star was faint to begin with and required a telescope with a high power eyepiece to see.

To take the photo, you need to be able to focus on the star and also to check that you are correctly tracking it as the photo is being taken.

There is a piece of equipment that can assist here, called a *radial guider* (shown next page).

The radial guider sits between the telescope and the camera and has a separate barrel that can fit onto your eyepiece, as well as the camera.

The barrel is adjustable to a point, which enables it to be positioned

where you need it for comfort.

Into the barrel goes your eyepiece for viewing and tracking. Some people also insert a Barlow lens between the eyepiece and barrel to increase magnification.

At the back of the barrel goes your camera body.

The radial guider's function is twofold. First, it allows you to attach the camera to the telescope. Second, it allows you to attach the eyepiece to the telescope.

Many people use a special eyepiece here called an *illuminated reticle*.

This eyepiece has double crosshairs, which is at the center of a boxed area. You position the star exactly inside the box, and, while shooting, check that the telescope is tracking properly.

The illuminated reticle is often used with

Camera on telescope with cable release

a Barlow lens to increase the image size, so that it fits snugly between the crosshairs and fills the boxed area.

If the image fits exactly in the boxed area, it is much easier to tell if your tracking is going off course.

There are other quite good methods of tackling the problem of tracking that goes off course, but space considerations prohibit going into them here.

Now, you can correctly focus and track the star that you wish to photograph.

Open the shutter and keep an eye on the image in the eyepiece.

The image in the eyepiece will be moving across its elliptical path.

So, as soon as you see it move across one of the crosshairs in the illuminated reticle, adjust the telescope to bring it back into position.

There is much much more to astrophotography but, hopefully, the short summation above will get you started and on the right track.

If you mess up, the stars will be there again tomorrow night.

One final thing with astrophotography using film where you are going to have it processed by a commercial film lab.

Since, unless you are shooting the Sun or Moon (remember the precautions here!) your images will be mainly dark with small dots of light.

A commercial film lab needs to see where the first frame is so it can set its equipment up properly to cut the film.

If the first frame is dark there will be a good chance that they will get it wrong and cut the film at the wrong position.

This will not make your day when you see the results so it can be a good idea to shoot the first frame of film you plan to use for astrophotography on a daylight subject which will give the lab a correct starting point.

You will find that you will be much happier also,

Happy shooting!

A radial guider, which allows the eyepiece and the camera to be connected to the telescope at the same time.

Binoculars: How They Work and What to Look For

Binoculars are simply two parallel refractor telescopes joined together.

Each individual telescope comes with a prism (with the exception of Galileo binoculars) in front of each eyepiece to correct the tendency of images produced by refractor telescopes to be mirror-reversed and upside down.

The two scopes are joined together, with a common adjustment control to focus both barrels at the same time.

A few models have separate adjustment controls for each barrel, which works fine for astronomical and marine viewing, where there is single focus distance.

These types are a little more robust than the binoculars with common focusing systems, but most people prefer the latter.

Specifications for binoculars vary from maker to maker and model to model, but are always expressed something like this: "10 x 50," "7 x 50," "6 x 30" and so on.

Simply, the first number refers to the binocular's magnification power and the second to the aperture size.

A pair of binoculars with 10 x 50 specifications has 10 x magnification power and an aperture size of 50mm.

If your interest is in astronomy, then you should consider binoculars as your first observing tool, before graduating to telescopes.

Many experienced astronomers recommend this and furthermore, they themselves use binoculars extensively, commenting that some of the most satisfying viewing they do is with binoculars.

Now why would this be so when binoculars have less magnification power than telescopes? Won't you see less?

Yes and no.

While binoculars will certainly not magnify the image as much as a telescope, they actually see more.

"Oh dear!" I hear you say, he's been sipping sherry in the cupboard again.

Not true, honest.

How can you see more with binoculars if they don't magnify an object to the same degree as a telescope?

It all has to do with what's known as *field of view*, which is discussed in more detail later in this chapter.

As mentioned elsewhere in this book (see Glossary), field of view refers to the width of the image given by a particular optical tool, such as binoculars, telescopes and camera lenses.

With binoculars, you will not have the same magnification power as a telescope, but your field of view will be much greater, so you will see more of the night sky.

See, I told you I wasn't sipping sherry.

Another factor to consider as a beginner to astronomy is that, at least initially, they are cheaper than telescopes.

Even a pair of high quality binoculars will cost less than a good quality telescope.

If you discover that astronomy is not your cup of tea, at least your binoculars can be used for a number of other applications.

You can take terrestrial photographs with some models, but as we will discuss further, it can be a bit cumbersome.

The type of binoculars you purchase depends on what applications you intend to use them for, in addition to some other a few other factors, which we will now discuss.

Binocular Types

The first factor to consider is what styles of binoculars are available. There are basically two styles available today, the Prism and Galileo binoculars.

The Prism binocular employs prisms to alter the light path, so that the image presented to the observer is not mirror-reversed and upside down.

This style of binoculars then divides into two sub-categories, called Porro Prism and Roof Prism.

As light passes through the barrels of Porro Prism binoculars, it is bent in a Z direction to ensure that the image is not upside down or mirror-reversed.

The Roof Prism design allows the construction of shorter and more compact binoculars by allowing the light to pass through the barrel in, virtually, a straight line.

At least, some manufacturers claim that it is a straight line, but in fact, this is a tad misleading.

Porro Prism binoculars, in which the eyepieces are on the same level as the objective lens.

You see, the light has to pass through two prisms to reach the eyepiece, and this certainly diverts the light part from the straight and narrow.

However, when the light exits the prism, it continues toward the eyepieces on the same horizontal, but not vertical, path.

Roof Prism binoculars, in which the eyepieces are set above the level of the objective lens.

Both styles use convex lenses at the objective lens and the eyepiece.

Please note, if you are using binoculars for astronomy, that you should only consider Porro Prism models, because as Roof Prism binoculars produce spikes of light when focused on bright stars, which can inhibit viewing.

Galileo binoculars are not suitable for extreme distances. They are more commonly used as opera and theater glasses.

This is because no prisms are used to correct the image. Instead, concave instead of convex lenses are used to produce an image that is not upside down or mirror-reversed.

Galileo binoculars, which use concave instead of convex lenses that restrict their effective range.

The use of concave lenses restricts the effective range of this style of binocular, though they are ideal for the theater or opera due to their compactness and light weight.

Magnification

Magnification is simply a way of expressing how big an object looks through binoculars, as opposed to the naked eye.

For instance, let's say your pair of binoculars has a 7x magnification factor

and you use it to look at an object from a distance of 350 meters.

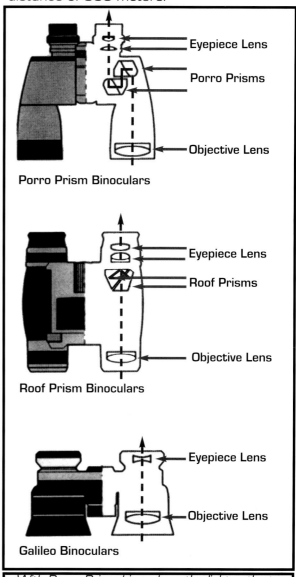

Porro Prism Binoculars

Roof Prism Binoculars

Galileo Binoculars

With Porro Prism binoculars, the light path stays on the same vertical level, but alters its horizontal plane at the prisms. The opposite applies to Roof Prism binoculars, which makes them shorter than Porro Prism. With Galileo binoculars, the light path is straight and does not diverge.

Because the binoculars magnify the image 7 times, the object will be seen as if you were viewing it with the naked eye at a distance of 50 meters.

This is calculated by dividing the distance to object (350 meters) by the binoculars' magnification factor (7), (350/7 = 50 meters).

Under the same circumstances, if you were using binoculars with 10x magnification, the object will be seen as if you were viewing it with the naked eye at a distance of 35 meters, (350/10 = 35 meters).

Binoculars directly change the resolving power of your eyes. Resolving power means how clearly your eyes can discern detail.

So a pair of binoculars with 7x magnification power increases the resolving power of your eyes by a factor of 7 times.

Please note that resolving power can be decreased by factors such as image shake and aberrations within the binoculars. This is covered later in this chapter. Other things also affect resolving power, such as humidity, heat haze and pollution.

It is also very important to remember, when discussing magnification, that the more you magnify a view the more any defects in viewing conditions or optical design will show up.

Additionally, to accommodate increased magnification, the image at the edge of the larger lens can deteriorate in quality.

Exit Pupil Size

Another factor to consider with binoculars is the exit pupil size.

The exit pupil size is the size of the round disc of light you see at the eyepiece, if you hold the binoculars up, just away from your eyes.

These two circles of light are actually the eyepieces' image of the light being transmitted by the objective lenses at the front of the binoculars.

The exit pupil size of a pair of binoculars is easily calculated by the following method. Take the aperture size of the objective lens and divide it by the magnification factor.

The two exit pupils (white discs) are clearly defined in this photograph.

For example, with a pair of binoculars with 10 x 50 specifications, you would divide the aperture of the objective lens (50mm) by the magnification factor (10) to arrive at an exit pupil size of 5mm (50/10 = 5mm).

If the binoculars has 7 x 50 specifications, then the exit pupil size would be 7.14mm (50/7 = 7.14). Easy isn't it? But is it important?

Yes, since the ratio between the exit pupil size of the binoculars and the size of the pupils in your eyes will affect the brightness of the image you observe when using them.

The pupils in the human eye vary in size, as the intensity of the light they are observing varies.

During the day, they are around 2-5mm in size. But at night, they increase in size to around 7mm to help compensate for the lack of light.

Now, if the exit pupils of binoculars are the same size, or larger, than your own pupils, then the brightness of the image you are observing through the binoculars will be the same as if you observed it with your own eyes.

The only difference the binoculars will make is to magnify the image.

This is the optimum scenario. Please note that the fact that the exit pupil size is bigger than your pupils does not mean that you will get a brighter image compared to using your eyes. The human eye compensates for the difference by reducing the size of the pupils.

However, if the exit pupil of the binoculars is smaller than your pupils, then the brightness of the image you are observing will be diminished by one factor.

That factor is the ratio of the exit pupil size in relation to your pupil size, at that particular time.

Let's see how this works with a pair of 10 x 50 binoculars. It's daytime and you are using these binoculars, which have an exit pupil size of 5mm (50/10 = 5). We have already said that during the day the pupil of the human eye dilates to between 2-5mm.

So the exit pupil of the binoculars is 5mm and your pupils are 2-5mm, which means the binoculars' exit pupils are bigger (or the same size) than your pupils, so the brightness of the image you are observing, in this scenario, will be the same as if using your own eyes.

Naturally, the magnification is increased due to the binoculars, but the brightness level is the same.

Now it night time and again you are using the same 10 x 50 binoculars with the 5mm exit pupil size to observe a scene.

Due to the reduced light, your pupils have now increased in size to 7mm, which is larger than the exit pupil size of 5mm.

The effect is to diminish the brightness of the scene you are observing by a factor.

By how much is the brightness diminished? It is simple to calculate.

The exit pupils of the binoculars are 5mm each and each pupil in your eye is 7mm, which expressed as a percentage is 71.43% (5/7 X 100 = 71.43%).

This indicates a loss of brightness, when using the binoculars in this example, of 28.57% (100% – 71.43% = 28.57%).

So, if night viewing is important, you need to carefully consider the magnification power of the binoculars you are planning to use and how that it will affect the exit pupil size of the binoculars.

If you use binoculars that have an exit

pupil size smaller than your eyes, then your view is reduced due to the lowering of the brightness of the image through the binoculars.

However, please note that with increased exit pupil size there is a price, and that price can be increased size, weight and cost.

You should also consider the fact that as humans get older the maximum size of the eyes pupils will diminish.

After around age 30–35, the pupils tend to only expand to around 6mm, and then further reduce as we age to a 4–5mm maximum at around ages 45–55.

This reduction in the flexibility of the pupils is more pronounced if you smoke.

So, a smaller binocular exit pupil size may not be so dramatic for those of us who have reached a more mature age (or have bad habits).

You should have binoculars that have exit pupils at least equal in size to your own pupils.

However, it is better if the binoculars' exit pupils are bigger than your eyes pupils.

This is because the human eye, when it observes a subject, does not sit still. Instead, it moves slightly from side to side, and this helps give us a three-dimensional image.

Using binoculars with exit pupils that are bigger than your pupils accommodates this movement, making viewing easier and less tiring.

Field of View (Real/True and Apparent)

At the start of this book, I promised to try and keep things simple and easy to understand.

So, I intend to explain this as simply as possible without employing scientific terms, in the hope that these concepts will be clear to you.

Binocular manufacturers often quote both real and apparent fields of view in their literature, and you can be forgiven for being totally confused as you read the literature provided.

For instance I have a brochure in front of me from a major manufacturer of binoculars.

It gives the specifications for all its models, and a closer inspection shows that the brochure specifies that their 8 x 23 model has the following fields of view: -

Real Field of View: 6.4° (degrees)
Apparent Field of View   51.2° (degrees)

Now, I think you would agree that there is a considerable difference between 6.4 and 51.2 degrees.

So is this manufacturer simply trying to confuse us? Not at all, but before we go into that I would ask you to conduct a little experiment.

Stand up and hold your left arm straight out in front of you. Now hold your right arm straight out to your side.

The arc between your right and left hands is approximately 90 degrees. So just over half of this distance is around 50 degrees.

Now move your right arm around to the approximate 50-degree position.

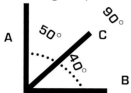

If you where standing at the intersection of lines A & C looking out, the vista you could see confined by these lines (lines A & B), would be roughly equivalent to what you can see with your eyes.

Look at the range between your two arms. It's quite wide and, of course, it increases as the distance increases.

If you have ever looked through a pair of binoculars, you will very quickly realize that there is no way any binoculars will produce a view that wide.

So what is this 51.2 degree apparent field of view that the manufacturer is talking about?

To sort that out, let's first deal with the real field of view.

Real Field of View

Real field of view refers to the width of the view you can see through a pair of binoculars, expressed as degrees of an arc, which starts at the point you are standing

As the arc stretches further out the width of the distance between its two constraining lines widens at any given point,

but is still exactly the same number of degrees.

Please note that even if the magnification rating of two pairs of binoculars is identical, the real field of view can be different between the two models.

This can be due to the optical structure of the binoculars being compared. In general, the real field of view decreases with increased magnification power.

For instance, a pair of binoculars with 7x magnification will have a greater real field of view compared to a pair with 10x magnification.

This is the same with all optics. The greater the magnification the less the real field of view.

As objects increase in size, due to magnification, the real field of view decreases.

With binoculars, the real field of view is often expressed (in the specifications) as so many degrees, or as a specific width at 1000 meters.

For instance, the Canon 15 x 50 IS AW binoculars specify the real field of view as being 4.5 degrees, or 79 meters (wide) at 1000 meters.

The specifications for the Canon 10 x 30 IS binoculars

The real field of view of binoculars is much narrower that the apparent field of view.

I'm sure we have all, at one time or another, held our hand out in front of us and noticed that it has a slight tremor.

Well, this tremor transfers itself to the binoculars you are using and causes the image to jump around a bit. How much the image jumps depends on how bad the tremor is.

No, a stiff glass of scotch will not help!

To counter this, certain binocular manufactures have introduced various image stabilization systems into some of their products.

Now, different binocular manufacturers employ various technologies to combat the image shake problem and they are refining the technology constantly.

These include the Gyro system (Fujinon), the Mechanical system (Zeiss) and the Canon electrical method.

All these systems are designed to vary the prisms slightly to account for binocular shake and are sometimes known as the Vari-Angle prism method.

Image Stabilized binoculars from Canon.
The button, when depressed, activates the image stabilization feature.

It is hard to say which is best. Some require batteries to operate, which can run out at the most inconvenient times, while others seem to have a problem with panning, which refers to following a moving subject in a lateral/horizontal direction.

Other types take around a minute to get up and running, which can be a real problem if your subject is fast-moving.

As with most things, you should try the various methods yourself and arrive at your own conclusions.

Whichever system you choose, you can be sure of one thing: Binoculars fitted with IS are much more expensive that those that are not. But that's the way of the world, isn't it?

Another way to tackle the image shake problem is to mount the binoculars on a tripod, which will keep the image rock steady. It's is also considerable cheaper than an Image Stabilization system.

So, make sure that your binoculars have a screw mount that will fit a tripod.

Any type of tripod can be used, from the common camera-type to the specialized models with mounts designed for sky viewing over an extended period.

These special models have a cantilevered mount that enables viewing the night sky very easy, even when you are viewing directly overhead.

All-Weather (AW) Features

If you tend to use your binoculars in adverse weather conditions such as rain, snow or dust, it is important that they be a good robust "All Weather" model, which suitably protect against the elements.

They should be sealed with "O" rings and have a cover over the body that is easy to grip, in any conditions.

These rather large binoculars have been built especially for astronomy. You view the night skies through eyepieces that are at the front, between the lenses.

They should also have a robust design, as people who tend to use these types of binoculars seem to be more prone to knocking them about a bit, due to the nature of their activities.

Please note that there is no such thing as "Impact-Proof" binoculars. Note that if the manufacturer claims the binoculars are "Water Resistant," that it does not mean they enjoy being submerged. Avoid water like the plague!

In fact, you might as well avoid so-called Water-Resistent models entirely. Get a full waterproof model. It is the only way to go.

But be aware that even the claim that a model is fully waterproof model is still conditional.

One example is the label "Fully Waterproof at 5 meters for one hour"—and that's one of the best models around!

The coatings on the front lenses give these binoculars their purple/blue color cast when subjected to light.

Binocular Uses and Applications

The uses and applications of binoculars are nearly endless, but include:

> Military use
> Commercial use (many and varied)
> Hunting
> Any type of spotting (trains, planes and birds, etc.)
> Marine applications
> Sports applications (horse racing, football, etc.)
> Terrestrial viewing
> Astronomical viewing
> Hiking, bushwalking and trekking
> Theater viewing (as opera glasses)

Correct Use of Binoculars

Using both eyes, look through the binoculars and adjust the barrel spacing to match the spacing of your own eyes. The barrels should swing to adjust the distance between them.

The binoculars have been correctly adjusted when you have the barrels positioned so the view from each one merges into one circle.

If they don't, you have a problem and they should be serviced, or not purchased in the first place.

If the binoculars' eyepiece (normally only one) has a diopter adjustment ring, you need to use it to adjust the eyepieces to your specific vision.

Remember that each individual has different eyesight.

Let's say that the diopter adjustment is on the right eyepiece (facing the binoculars). Now close your right eye and, using the focusing knob, adjust the focus on the object being viewed so that is in focus in the left barrel when you view it using only your left eye.

Now close your left eye and look through the right barrel. Adjust the right diopter adjustment ring to bring it into focus.

If you wear glasses when using binoculars, the scene may fade towards the outer edges. To counter this, use the rubber eyepieces and diopter adjustment, if provided. Diopter adjustment controls are built into the eyepieces of better quality binoculars as are fold out rubber eyepiece covers.

If your binoculars require batteries to run their image stabilization system, always carry spare batteries.

Checklist for Purchasing Binoculars

It is very important that you think about what you are going to use the binoculars for before, not after, the purchase. Then, use the checklist below to help arrive at a sensible decision.

When purchasing binoculars, look through a variety of models and compare them. Always look at exactly the same scene, so you have a valid comparison of how each model performs.

You should look at the alignment of both barrels. They need to be absolutely parallel and any manufacturing defect, or damage during shipping, can knock this alignment out. Normally, it's the prisms that are bumped loose.

You can tell quite easily because if you look through them, you will always have two images and will be unable to adjust them to form one image.

If this happens to you, don't purchase them. Find another pair or brand that does not exhibit this problem.

Check that the lenses are coated and the potential light loss problem is corrected. Use the method described in the section on lens coatings earlier.

Focus on a scene that includes lettering, perhaps a street sign. Is the lettering crystal clear? It should be.

Now move the binoculars slightly, so that the lettering is at the edge of the image circle. Is it still crystal clear?

If not, then the binoculars could suffer from what's know as chromatic aberration, which is a defect in the lens design affecting the resolution of the image at the edge of the image circle.

Now a lot of people think this is not important as it only effects the edges of a scene and they tend to concentrate on the center of the image circle.

Not so, I'm afraid, as binoculars with this defect can be tiring to use and even induce headaches. Avoid this defect!

Don't purchase more power than you need. Buying like that just wastes money for power that you will never need, or use. And binoculars with more power are bulkier.

Are your hands steady? If not, and you need a very steady image, then perhaps you should be thinking about binoculars fitted with an Image Stabilization (IS) system or built to fit onto a tripod.

Will you use the binoculars in snow, rain or dusty conditions? If so, you need a good, reliable "All Weather" design.

If you are going to use them for night viewing, remember what we said about the exit pupil size of the binoculars. It might pay you to study that section more closely.

Compare prices and ensure you are getting value of money.

Do the binoculars come with a neck strap and lens covers? These minor accessories are important, first for ease of use and security (neck strap), and second, to help keep them in tip-top shape when not in use (lens covers).

Have a good look at the exit pupils. Are they perfectly round, with white light fully and evenly filling the white circle? Or do they have gray edges that form a pure white box within the circle?

If the box is there, it means the manufacturer has been a bit mean and skimped on the construction of the prisms. Avoid binoculars with this trait.

Also if there is any white light outside the circle, normally shaped like a wedge, move on to another pair of binoculars.

Lastly, grab hold of one of the eyepieces and try and pull it away from the binoculars.

If the arm that connects it to the center post and the opposite eyepiece flexes, then again the manufacturer has not produced the best product. Find binoculars which don't flex in this area.

Binoculars: Care and Maintenance

The care and maintenance of binoculars is fairly minimal, but a few points should be remembered.

Avoid letting the binoculars suffer bumps and hitting things. It's not good for their well-being!

Don't store binoculars in a car or another place where excessive heat or humidity can build up. Always store them with a sachet of silica gel.

When not used, keep the lens covers on to protect the objective lenses and eye-pieces. This protects them from scratches.

Make sure the binoculars are dry before using the lens caps to avoid fungus.

The same advice applies if the binoculars are very hot or cold, and are moved into an environment with normal temperature. Let their temperature stabilize before putting on the lens caps.

Always, when using binoculars have the neck strap where it is supposed to be, around your neck. The sound of binoculars slipping from your hand and striking a rock, or some other nasty hard object, is not a pleasant one.

Even worse is the splash you hear as the binoculars sink to the depths of an ocean or river.

When they get wet or dusty, clean them with a soft cloth. Avoid harsh solvents at all times and only, if necessary, use a mild, diluted detergent to clean them.

Keep the lenses clean using only lens tissues and lens cleaning solution, which are readily available from camera stores. You can also use a lens blower to remove dust.

Do not pour the lens cleaner directly on to the lens surface. It may get inside and ruin your binoculars. Instead, pour a little of the lens cleaner onto a lens tissue and wipe the lens in a straight lines, starting at the center and working towards the edges.

Once that is done, get a dry lens tissue and wipe in the same manner to remove any moisture and dirt. Do not use handkerchiefs and normal tissues to clean lenses, unless the idea of scratches across your lens appeals.

Also don't use the same lens tissue twice. What's the point of wiping dust and grit off a lens and then reapplying it to the lens surface?

If your binoculars use batteries to run their image stabilization system, remove the batteries and store them separately when the binoculars are not in use, in case they leak.

Under no circumstances attempt to disassemble the binoculars.

If there is a problem with your binoculars take them to get repaired. Get a quote before any work is carried out. You may find it's cheaper to replace, rather than repair.

Avoid touching the lenses with your fingers. Greasy fingerprints do not enhance image quality.

Glossary of Terms

Absolute Magnitude
A means of describing an object's intrinsic brightness, eliminating the effect of distance on its magnitude by calculating its brightness as if it lay at a fixed distance of 32.6 light-years from the Sun.

Achromatic Lens
Lens design where there is the use of two elements in the lens (positive and negative elements) that brings together two wavelengths of light to a common focal point.

Altazimuth Mount
A telescope mount that attaches the telescope to a tripod and allows movement in two directions: (1) horizontally and (2) vertically. The altazimuth mount does not allow horizontal polar movement with one control, as does the equatorial mount.

Annulus
A ring of fire around the Sun that is seen during a Solar Eclipse when, due to the orbit of the earth or the Moon, the Moon does not fully obscure the face of the Sun.

Aperture
With telescopes, *aperture* refers to the diameter of the objective lens or primary mirror and can be specified either in inches or millimeters. So, a telescope described as a 4-inch refractor will have an objective lens with a diameter of 4 inches, while one described as an 8-inch Newtonian will have a primary mirror of 8 inches. (Also see *Focal Ratio.*)

With cameras, *aperture* refers to the amount of light the lens allows to pass to the film plane of the camera and is normally controlled by an adjustable control, which can increase, or decrease, the aperture size. It is defined by a system of f-stops, for example f4, f5.6, f8 and so on. (Also see *f-stop.*)

Apochromatic Lens
A lens design that eliminates chromatic aberrations by use of different varieties of specialized glass, with low or anomalous dispersion. Special coatings of the lens are also used, including fluorite.

Apparant Magnitude
How large an object looks in the sky when viewed from Earth, based on the light it emits.

Astronomical Unit (AU)
A measurement based on the earth's distance from the Sun (92,752,000 miles equals 1 AU) that is used in astronomy. It is always expressed as distance from the Sun.

Astrophotography
The photography of object in space, usually with a camera attached to a telescope, but other means may be employed.

Barlow Lens
An extension tube for a telescope which, when used in conjunction with an eyepiece, will increase the magnification of the image being viewed by a factor expressed as 2x, 3x and so on.

Bright Star Atlas
A periodical that shows the position of heavenly bodies outside our solar system for a particular time period (epoch).

Cassini Division
The dark gap between rings A & B of Saturn.

Catadioptric Telescopes
An evolution of the catoptric design that employs a special corrector lens at the front of the tube, as well as a primary and secondary mirror.There are two variants know as the Schmidt-Cassegrain and the Maksutov-Cassegrain models.

Catoptric Telescopes
A telescope design that employs two mirrors (primary and secondary) to deliver the image at the focal point (ie. the eyepiece). They are also known as reflector and Newtonian reflector telescopes.

CCD
CCD stand for *charge-coupled device* and is a type of imaging sensor used in digital cameras.

Celestial Equator
The division of the skies into northern and southern hemispheres by the projection of a circle into space that is 90° (degrees) from either the North or South Celestial Poles.

Celestial Poles (North & South)
The points in the sky above the northern and southern hemispheres around which the stars appear to rotate.

Chromatic Aberration
A condition that causes images to blur due to the fact that the different wavelengths of light are focusing on slightly different focal points. It is corrected by the use of special glass and coatings in lens and mirror design.

Clock Drive
See *Motor Drive.*

CMOS
CMOS stand for *complimentary metal-oxide semiconductor* and is a type of imaging sensor used in digital cameras.

Collimation
To correctly align the optics of a telescope.

Compact Camera
A camera with a lens that is usually not removable. The photographer views the scene through a separate viewfinder,

Declination (Dec)
A system of measurement that measures how far north or south of the Celestial Equator an object is. Any object on the Celestial Equator has a Declination (Dec) of 0. Declination (Dec) is measured in degrees, minutes and seconds of arc.

Depth of Field
Depth of field refers to the distance, in front of and behind your point of focus, which is in sharp focus. It varies with aperture and lens selection. Lenses with a short focal length have greater depth

of field than lenses with a long focal length. In addition, apertures that let in less light (high f-stop numbers) have greater depth of field than apertures that let in more light (low f-stop numbers).

Dioptric Telescope
See *Refractor*.

Dobsonian Telescope
A reflector telescope on a special Dobsonian mount.

Ecliptic
The great circle representing the apparent annual path of the Sun; or the plane of the earth's orbit around the Sun. All of the planets rotate the Sun in approximately the same ecliptic.

Ephemeris
A publication that gives various information about the position of objects in our Solar System over a 12-month period.

Equatorial Mount
A device that attaches a telescope to a tripod and allows for the movement of the telescope horizontally (side to side), vertically (up and down) and also laterally along a circular path that mimics the curvature of the earth (right ascension). In addition, movements for declination are possible. Also known as German Equatorial Mount.

Equatorial Orbit
An orbit of Earth that follows the equator.

Equinox
Occurs when the Sun passes over the plane of a planet, which has the effect of making night and day equal in length.

Field of View
Field of view refers to the extent of view a particular optical instrument provides. It can be measured in degrees or in feet/meters at a particular distance. For instance, a five-degree field of view (say, using binoculars) means that you can see anything in a five-degree arc. In terms of distance, a 5°(degree) field of view means that at a distance of 1000 yards, your image would have a width of 262 feet.

Film Plane
The position in a camera where the film or image sensor resides.

Finderscope
A finderscope is a small, low powered, auxiliary telescope with a much wider field of view than your main telescope, used to help you find objects more easily.

Fluorite
A coating used on high quality lens to assist in combating chromatic aberration.

Focal Length
The focal length of a telescope refers to the distance between the primary mirror or objective lens and the point where the image is in focus, or the focal point. If you know the focal ratio and aperture size of your telescope, you can easily calculate the focal length. Simply multiple the aperture size (in mm) by the focal ratio. For example, let's say you have a telescope with an aperture of 4 inch-

es (101.6mm) and a focal ratio of f10. The focal length of this telescope is 1016mm (101.6 x 10 = 1016mm).

Focal Point
The point in an optical system where the light being reflected by the lens or mirror is in focus.

Focal Ratio
This refers to the f-stop or photographic speed of a particular telescope. (See *f-stop*.)
With telescopes, the focal ratio of a particular model is calculated in the following manner: Take the focal length of the telescope and divide it by the aperture size. For example, a telescope with a focal length of 1016mm and an aperture size of 4 inches, or 101.6mm, has a focal ratio of f10 (1016/101.6 = 10).

f-number
See *f-stop*.

f-stop
A system devised to express the light gathering power of a lens, normally expressed as f-4, f 5.6, f8 and so on. The higher the f-stop number, the less the light gathering power of the lens. The lower the f-stop number, the more the light gathering power. Each increase in the f-stop number represents a halving of the light gathering ability of the lens. For example, a lens rated at f11 will gather half as much light as one that is rated at f8. This is the same for telescope and camera lenses.

Inner Solar System
Includes the Sun, Mercury, Mars, Earth and Venus.

ISO
ISO stands for *International Standards Organization* and refers to a film or CCD sensor's sensitivity to light. Film is manufactured with various ISO ratings and sensors in a digital camera may be set at various ISO rates.

A low ISO (ISO 25-100) number means low sensitivity and a high ISO number (ISO 100 +) indicates a greater sensitivity to light. When very high ISO settings are used the exposure time for astrophotography is shortened, but there is a trade-off. With high ISO rated film (around ISO 400 +), images can start to appear grainy, and with CCD sensors, computer noise can be generated within the image, which gives a similar effect to grain (around ISO 800 +).

Geosynchronous Orbit
An orbit of a satellite that matches the planet's rotation, thus keeping it at one point above the planet.

Lunar Eclipse
A Lunar Eclipse occurs when the earth comes between the Moon and the Sun.

Maksutov-Cassegrain Telescopes
See *Catadioptric Telescopes*.

Main Belt
An area between Jupiter and Mars where there are many thousands, perhaps millions, of asteroids.

Moon Phases

There are eight phases of the Moon, which range from Full Moon to New Moon. Each phase describes how much of the Moon's surface can be seen from Earth at that time.

Motor Drive

A system of motors and gears that is used in conjunction with telescope mounts to move the telescope in synchronization with the rotation of the earth. It is especially helpful in astrophotography.

Newtonian Reflector Telescopes

See *Catoptric Telescopes*.

Opposition

The alignment of two celestial bodies on opposite sides of the sky as viewed from the earth. Opposition of the Moon or planets is often determined in reference to the Sun.

Outer Solar System

Jupiter, Saturn, Uranus Neptune and Pluto.

Planisphere A device for identifying what heavenly objects are showing in the night skies on any particular date.

Penumbra

See *Solar Eclipse*.

Polar Orbit

An orbit that circumnavigates Earth by passing over both North and South Poles.

Prime Meridian

An imaginary line which runs due north/south through the vernal equinox, which is the point on the Celestial Equator where the Sun crosses on March 20th each year. The Prime Meridian runs through Greenwich, England, and is the basis of the division of the earth by longitude.

Proper Motion

The shift of a stars position in the heavens which is measured in arcseconds.

Radial Guider

A device allows the attaching of both a camera and the eyepiece to a telescope, at the same time.

Rangefinder Camera

See *Compact Camera*.

Right Ascension (RA)

A system used to measure how far east of the prime meridian an object lies. Right Ascension (RA) is given in hours, minutes and seconds of time and is measured in an eastwardly direction from the Prime Meridian which has a RA of 0 hours.

Reflector Telescopes

See *Catoptric Telescopes*.

Schmidt-Cassegrain Telescopes

See *Catadioptric Telescopes*.

Setting Circles

Setting circles are dials on the mount that are engraved and show Right Ascension (RA) and Declination (Dec).

Shutter Speed

The time that a cameras shutter is open allowing light to strike the film or imaging sensor.

Single Lens Reflex (SLR) camera.

A camera that allows the photographer to view the scene through light that has been passed through the camera lens, via a pentaprism. Usually, its lenses are interchangeable.

Spotting Scope

A specialized refractor telescope.

Solar Eclipse

Occurs when a Full Moon come between the Sun and the Earth. The area on Earth that the shadow is cast upon is either full (Umbra) or partial (Penumbra).

Star Diagonals

The design of some telescopes dictates that the eyepiece is positioned down at the end of the tube which can make viewing (especially the sky) both difficult and uncomfortable. A star diagonal, which is a prism or mirror set at 90 degrees to the telescope tube, allows placement of the eyepiece in a much more comfortable viewing position.

Terminator

The line that defines where light finishes and shadow starts on the Moons surface when viewed from Earth.

Transit

When one planet passes over the face of of the Sun

Umbra

See *Solar Eclipse*.

Index

Acknowledgments

Trying to write a simple book on what to newcomers appears complicated, is no easy task.

I would not have been able to complete this book, at least in the manner I wished, without some major assistance from a number of individuals and organisations.

Firstly I have to thank Tasco Sales (Aust) Pty Ltd and their sales manager Kevin Johnson for their support and assistance especially in regard to images, illustrations and technical advice.

In particular my heartfelt thanks to Ian James, Tasco's technical genius, for his unflagging assistance throughout this project. I don't think I could have finished, on time, without it.

Again for technical help, photographs and illustrations I also need to thank Ian Threasher at Canon Australia Pty Ltd and his staff.

Canon

Steve Atkinson and his staff at ExtraVision Pty Ltd also deserve a big thanks for technical assistance, photographs and support over the last nine months.

Additional much need support was received from Mick Smith and the staff of The Binocular and Telescope Shop and Michael Weston and Shay Mclean from York Optical Pty Ltd, both in Sydney.

A special thanks to Les Sara from The Astronomical Society of NSW for his guidance, assistance and especially his patience of which, fortunately for me, he had a plentiful supply.

As always my friend Mark Queenan from Fletchers Fotographics was a source of infinite assistance and patience.

Any omissions or mistakes in this book are of my doing and do not reflect on the advice and assistance these kind people have given me.

Last but not least my wife Jenny has been right beside me throughout this project, although I'm sure, at times, there were other places she would have preferred to be.

I can't thank her enough but I'm sure she will think of some way.

Bill Corbett
August 2002